"A gripping and absolutely terrifying blow-by-blow account of the way that companies, governments, cops and crooks have entered into an accidental conspiracy to poison our collective digital water supply in ways small and large. . . . An incandescent call to arms demanding that states and their agents . . . join with civil society groups that work to make the networked society into a freer, better place than the world it has overwritten." – Cory Doctorow, *Globe and Mail*

"*Black Code* effortlessly chronicles threats ranging from individual privacy to national security ... [and] highlights the shadowy, lucrative war online, behind closed doors and in the halls of power, which threatens to control, censor, and spy on us, or worse." – *National Post*

"Ron Deibert is an excellent guide to the fascinating and disturbing world of cyber security." – Joseph S. Nye, Jr., Distinguished Service Professor, Harvard, and author of *The Future of Power*

"At a time when autocrats, criminal gangs and others are trying to control and pervert the use of cyberspace, Ron Deibert's *Black Code* rings like a fire-bell in the night, warning us that the price of a new global commons of shared knowledge and connectivity is vigilance in defense of free expression and the rule of law." – Carl Gershman, President, National Endowment for Democracy

"*Black Code* stimulated my thinking about the potential for making the Internet a much safer place." – Vint Cerf, Internet Pioneer

"A must-read for Netizens!" – Margaret Atwood, on Twitter

"A timely book . . . [Its] essential political message is as old as Toqueville, and more vital than ever." – *Quill & Quire*

Ronald J. Deibert is professor of Political Science and Director of the Canada Centre for Global Security Studies and the Citizen Lab at the Munk School of Global Affairs, University of Toronto, an interdisciplinary research and development "hothouse" working at the intersection of the Internet, global security, and human rights. He is a co-founder and a principal investigator of the OpenNet Initiative and the Information Warfare Monitor, which uncovered the GhostNet cyberespionage network of over 2,500 infected computers in 103 countries. Deibert's work has received front-page coverage in the *Toronto Star, Globe and Mail, International Herald Tribune*, and *New York Times*. He lives in Toronto with his family.

BLACK
CODE

SURVEILLANCE, PRIVACY, AND
THE DARK SIDE OF THE INTERNET

RONALD J. DEIBERT

SIGNAL
McCLELLAND
& STEWART

For Joan

Signal is an imprint of McClelland & Stewart,
a division of Random House of Canada Limited, a Penguin Random House Company.

Library and Archives Canada Cataloguing in Publication is available upon request.

ISBN 978-0-7710-2535-8

Library of Congress Control number: 2013938866

Typeset in Bembo by Erin Cooper
Printed and bound in the United States of America

McClelland & Stewart,
a division of Random House of Canada Limited,
a Penguin Random House Company
www.penguinrandomhouse.ca

2 3 4 5   18 17 16 15

Penguin
Random
House

CONTENTS

AUTHOR'S NOTE TO THE PAPERBACK EDITION

Who are we? Where do we stand in relation to each other, and more importantly, to the state? These are timeless questions worthy of constant re-examination, never more so than today in our world of Big Data, social networks, and global communications. According to the U.S. National Security Agency (NSA), we are what we communicate. What rights as citizens do we have to these communications being private? Again, according to the NSA, none.

In May 2013, as the hardcover version of *Black Code* was being prepared for publication, thousands of miles away in Hawaii, a private contractor for the National Security Agency named Edward Snowden was busy making preparations for his escape from home and what was to become the most extensive single set of leaks to ever hit the U.S. intelligence community.

There are many interesting details about Edward Snowden and his access to top-secret programs. The twenty-six-year-old network systems administrator worked for only three months at Booz Allen Hamilton, one of the many private companies that orbit "Top Secret America," the Cyber Security Military Industrial Complex. A systems administrator, or *sysadmin*, is the type of person you might call to fiddle with the cables under your desk when your Internet connection is not working. And yet, notwithstanding his short period of employment and junior status, Snowden claimed to have the ability "to wiretap anyone, from you or your accountant, to a federal judge or even the president, if I had a personal email." U.S. officials disputed this as wildly distorted, but documents Snowden later revealed to the *The Guardian*, specifically a PowerPoint presentation that gives an overview of a classified Internet monitoring program called XKeyscore, support it. That the NSA can monitor anyone's Internet activity, and seemingly everything they do online in real-time, is shocking in and of itself; that a twenty-six-year-old sysadmin working only three months for an outside private contractor could do so is heart-stopping.

Much has been made about Snowden's motivation. His choice to flee first to Hong Kong, and then to Russia, where he spent weeks in an airport transit lounge before being granted a temporary one-year asylum in that country, left many to suspect he is a mole for a U.S. adversary. Some have suggested that he took the position with Booz Allen Hamilton simply to get access to classified materials and then leak them to reporters and WikiLeaks, and was perhaps even encouraged by WikiLeaks to do so. Many believe he is a traitor, not a whistleblower – that he has done enormous harm to the U.S government while aiding adversaries like China, Russia, and even al-Qaeda.

As intriguing as it may be to speculate about Snowden's motivation, about Snowden the man, such speculation distracts from the details of the leaks themselves. The "Snowden Files" have blown wide open intricate details about programs that operate deep in the shadows of the classified world, hidden from not only most citizens, but from lawmakers too. At the time of writing, it is believed that Snowden has released only a small fraction of the material that he acquired before fleeing Hawaii: more is likely to come in the weeks and months ahead. What has already been published is remarkable enough for what it reveals about the extent of eavesdropping on our digital lives and the collusion of some of our most familiar and trusted Internet brands with secretive and largely unaccountable security forces.

Among the revelations:

- ••• The U.S. government ordered Verizon, a top-tier global telecommunications company, to provide it with access to the "metadata" for all of the communications made through its service, including metadata for domestic communications, a violation of the NSA's mandate and possibly the U.S. Constitution. What is metadata? Metadata is the electronic wrapper that accompanies every digital communication. For example, my mobile phone, even when I'm not using it, emits an electronic pulse every few seconds to the nearest wifi router or cellphone tower that includes a digital biometric tag: the model of the phone, its operating system, the geolocation of the phone (and by extension all of my movements). Meanwhile, when the phone is in use metadata includes the number I'm calling, the length and time of the call, or the IP addresses of websites I visit. All of this metadata moves through the filters and chokepoints of the Internet, and sits indefinitely, there to be mined, on the servers of the companies that own and operate the infrastructure, such as telecommunications and Internet service providers like Verizon.
- ••• Microsoft, the world's largest software maker, helped the NSA and FBI collect data on users of its products and services and to access the encrypted chats and conference calls made over Skype, a service that Microsoft purchased in 2011. Microsoft also opened up its cloud computing services and email products to U.S. security agencies.
- ••• Through a program codenamed PRISM, the U.S. government required Google, Microsoft, Yahoo!, Facebook, Apple, and several other companies to facilitate direct

access to customer data managed by the companies and compelled the companies to remain silent about these arrangements under penalty of law.

••• The NSA and GCHQ worked systematically to defeat encryption by having tech companies insert secret "back doors" and by covertly influencing international organizations to adopt weak encryption protocols.

One of the more remarkable issues to emerge from these revelations is the ineffectiveness of oversight mechanisms, and sometimes their outright perversion. For example, the very court set up to be a check against precisely this type of surveillance, the U.S. Foreign Intelligence Service Court (FISC), has not only allowed it, it has created a body of law that condones the mass collection of U.S. citizens' communications. The FISC operates in secret, its deliberations and decisions shielded from public scrutiny. It has overseen thousands of requests for approval of sweeping collection programs during President Obama's administration, and it has denied only one.

The leaks have also laid bare how semantic sleights of hand have become customary among those accountable, including the President of the United States, who said coyly when referring to the collection of metadata through the Verizon-NSA arrangement: "no one is listening in on your phone calls." Technically true, maybe . . . if we can take Obama at his word. But they *are* keeping a record of every time you make a call and to whom the call is made, how long the call lasts, and where each party to the call is physically located, and they have been doing this for at least seven years. Across the board, government officials employ linguistic evasion about programs like this without shame or fear of consequence. James Clapper, the Director of National Intelligence, flat out lied to Congress when he responded "no" when asked whether the NSA collects American communications. Called out on the claim, he later admitted it was the "least untruthful" answer he could give. Consider the NSA's idiosyncratic understanding of the term "collection." As the NSA vacuums up every bit and byte of digital information, it has redefined "collection" to mean that which happens only when an analyst starts interrogating the data, how should we say it, that has been *acquired*? In a similar vein, Section 215 of the U.S. Patriot Act allows the government to compel businesses to disclose information "relevant" to foreign intelligence investigations, and the government has used this capacity to collect all phone records because, well, because they are *relevant*. The same semantic contortions have been applied to terms like "incidental," "inadvertent," and "minimize." In an episode of *Seinfeld*, George explains to Jerry, who is interested in finding out how to evade a lie detector, "just remember, it's not a lie . . . if you believe it." For the U.S. government, it seems such words of comedic wisdom have become public policy.

Behind these revelations is an apparent sea-change in the modus operandi of the NSA that can be traced back to 9/11 and the perceived failure to connect the dots leading up to those horrific attacks. Insiders have described how the NSA's chief, General Keith Alexander, prompted a shift in the NSA's methodology from a targeted surveillance approach to one that vacuums up and stores all human communications for later retrieval and analysis. According to the *Washington Post*:

"'Rather than look for a single needle in the haystack, [Alexander's] approach was, *'Let's collect the whole haystack,'* said [a] former senior US intelligence official who tracked the plan's implementation. *'Collect it all, tag it, store it. . . . And whatever it is you want, go searching for it. . . . '*"

About such programs many insist that if you have nothing to hide, you have nothing to fear. This trivializes what's at stake. Checks and balances are at the heart of liberal democracy for a reason: to prevent the abuse of power. Only a few decades ago, U.S. president Richard Nixon used the machinery of government, including its law enforcement and intelligence agencies, to cover up crimes for which he was responsible, and to harass and intimidate those with whom he disagreed. For anyone who feels that unchecked surveillance programs can do no harm, the Watergate episode should be required study. We are talking about a substantial structural flaw at the heart of government: we cannot permit secret arms of the state to indiscriminately listen in on, watch, or otherwise collect everything we do and say online, and still call it liberal democracy.

• • •

It is hard to say what lasting impact the Snowden revelations will have but they appear to be cutting in at least two directions. First, they are at least momentarily spurring greater public curiosity about what happens beneath the surface of our digital world. Each Snowden revelation has fuelled journalists and others to probe, dig deeper, ask more questions. Citizens in countries other than the United States are beginning to find out about their national counterparts to the NSA, and what type of international information sharing has gone on among intelligence agencies worldwide. Whether this type of active curiosity will continue through short-term news cycles and information overload, however, is an open question.

The second impact is less positive, and more ominous for our digital future. A huge proportion of global Internet traffic flows through networks controlled by the United States, simply because eight of the fifteen global tier 1 telecommunications companies are American – companies like AT&T, CenturyLink, XO Communications, and Verizon. The social media services that many of us take for granted – including Google, Facebook, Yahoo!, and Twitter – are also mostly provided by giant corporations head-quartered in the United States. All of these companies are subject to U.S. law, including the U.S. Patriot Act, no matter where their services are offered or their servers located. Having the world's Internet traffic routed through the U.S., and having those companies under U.S. jurisdiction, gives U.S. national security agencies an enormous home-field advantage. This awareness will undoubtedly trigger a reaction abroad as policymakers and ordinary users realize the huge disadvantages of their dependence on U.S.-controlled networks in social media, cloud computing, and tele-communications, and become aware of the formidable resources that are deployed by U.S. national security agencies to mine and monitor those networks. Already we can see regional traffic to the United States from Asia, Africa, and Latin America gradually declining, a trend certain to accelerate as policymakers ramp up regional network exchange points and local services to minimize dependence on networks under U.S. control. Many of the countries in the Southern Hemisphere are failed or fragile states; many are authoritarian or autocratic. No doubt elites there will use the excuse of security to adopt more stringent state controls over the Internet in their jurisdictions and support local versions of popular social media companies over which they can exact their own nationalized controls – a trend that began prior to the NSA revelations but which now has additional rhetorical support. Many will no doubt use the threat of the NSA to convince other countries to join them to wrest the Internet from "U.S. control" but in practice will subject it to their own secretive programs of censorship and surveillance. The most profound impact of the Snowden revelations may be one that comes back to haunt us all in the long run: the end of an open Internet.

• • •

For me and many of my colleagues who study Internet security issues, the Snowden revelations have not come as a real surprise. We may not have known the specific codenames or the full scope of what was involved, but we had a good idea of the contours. Even many in the general public have long suspected something like what Snowden brought to light, hinted at in popular culture productions like the *Bourne* series. And, to be sure, the world can be a nasty place. We need law enforcement, defence, and national intelligence agencies to do their job. Only the most naïve optimist would advocate for their elimination.

But how many of us knowingly consented to this type of whole-sale Big Brother surveillance when we signed up to our Internet's terms of service contract? Did we click "I agree" with an understanding that our lives would be reduced to data points exposed to mass collection? In the world of Big Data, where we are turning our digital lives inside out, should we be entrusting power and authority to agencies that barely acknowledge their own existence, that operate outside the rule of law?

PREFACE

It always takes long to come to what you have to say, you have to sweep this stretch of land up around your feet and point to the signs, pleat whole histories with pins in your mouth and guess at the fall of words.

— Dionne Brand, "Land to Light On"

May 24, 2012. Calgary, Alberta. I am at a cyber security conference with the disarming title "Nobody Knows Anything." In attendance are academics, private sector representatives, and senior government officials. *Surely these people know something*, I think to myself. Perhaps not. All Canadians have heard of the Royal Canadian Mounted Police (RCMP), and most the Canadian Security Intelligence Service (CSIS), but stop a random sample on, say, Yonge Street in Toronto, and ask if they've ever heard of the Communications Security Establishment Canada (CSEC) and most will shrug. This is because CSEC, Canada's version of the U.S. National Security Agency (NSA), is the most secretive intelligence agency in the country. *Nobody Knows Anything,* I think. *How convenient.*

I am on a panel with John Adams, the recently retired chief of CSEC and once Canada's top spy, and Harvey Rishikof, an American lawyer, and now a professor at the National Defense University in Washington, D.C. Rishikof has had a distinguished career in national security, and at various times was the senior policy advisor and legal counsel to the FBI and the Office of National Counter-intelligence Executive (NCIX) at the Directorate of National Intelligence (DNI). I felt lost.

When my turn to speak comes around I joke about the title of the event. I explain that I was a little confused by it at first, but upon reflection and after looking at the roster of spooks, ex-spooks, and wannabe-spooks in attendance, it suddenly all made sense. *Nobody Knows Anything.* "Of course," I say, "this is all about plausible deniability!" I had forgotten the first rule of public speaking: Know your audience and tell them what they want to hear. *This was not going to go well*, I thought to myself – and it didn't.

When his turn comes, Rishikof brings up the Citizen Lab's *Tracking GhostNet* report in positive terms, but then demurs. "We [U.S. national intelligence agencies] would not have been able to do what Deibert and his group did with the GhostNet investigation," he says. "Trespassing and violating computers in foreign jurisdictions . . ."

Trespassing? Violating computers in foreign jurisdictions!

Here we go again, I say to myself, and in rebuttal, attempt to dispel misconceptions. I insist that the Citizen Lab did not trespass or violate anything, and certainly not "computers in foreign jurisdictions." We simply browsed computers already connected to the public Internet, and did not force our way into them. Rather, the computers were configured (by their owners) in such a way that their contents were openly displayed to us (and to anyone else who made the effort). Sure, the attackers may have erred by serving up content that they didn't want others to see, but the bottom line was that they offered up information to anyone who connected to those computers. We just knew where to look. If this is trespassing then so is just about everything that happens online.

As the panel ends, we pack up our material and exchange pleasantries. Adams walks over to me and says in a grave tone: "You know, Ron, there were some people in government who argued that you should be arrested." Grinning broadly, he laughs. "And I agreed with them!"

Over the last decade, there have been many times like this when I have wondered, as the Talking Heads put it, "How did I get here?"

• • •

They were heady days. It was spring 2001, and I had just received authorization to set up the Citizen Lab at the University of Toronto. The initial funding came from the Ford Foundation, and the idea was simple: To study and explore cyberspace (though few called it that back then) in the context of international security. The dot-com era was in full swing, the Internet and "information super-highway" spreading like a brushfire, timeworn political divisions – the Cold War, South African apartheid, and so on – relegated to history books, and generally people were in a good mood, a very good mood. At the dawn of the twenty-first century it was hard not to be an optimist.

9/11 ripped into all of that and left us all reeling, for the next year or so most of us wondering what kind of world do we now live in? In January 2003, I published an article in *Millennium*, a journal published by the London School of Economics, arguing that this singular event had reshuffled the deck around issues relating to cyberspace, and that trouble was brewing. Rightly or wrongly, those planes smashing into New York's World Trade Center, the Pentagon, and a field in Pennsylvania were viewed as a failure of cyber intelligence, of authorities not monitoring Internet commu-nications and activities closely enough. At the same time, the pre-vailing view for most of those connected was that the Internet could not be controlled by governments: "The Net interprets censorship as damage and routes around it," as John Gilmore, founder of the Electronic Frontier Foundation, once famously quipped. I was not so sanguine. That article has been haunting me for years; it only touched the surface, and has struck me ever since as unfinished business. It was called "Black Code."

National security apparatuses have deeply entrenched, subterranean roots whose spread is difficult to curtail, let alone reverse. When there is human agency involved – while the Internet often seems to be operating in an ethereal realm, it has proven itself human, perhaps all-too-human – those responsible for security rarely agree that something is outside their control. Instead, they ramp up. Some governments in the 1990s were already erecting borders in cyberspace, long before 9/11 shifted the terrain around state surveillance and gave it added impetus. Anti-terrorism laws unthinkable on September 10, 2001 were proclaimed with little public debate across the industrialized world, and the United States in particular (but certainly not alone) began quietly building offensive cyber attack capabilities. The enemy was terrorism, an abstract noun, but al-Qaeda was a real and immediate foe. I wrote in a *Globe and Mail* op-ed on January 1, 2003: "Government armed forces from around the world have devoted increasing time, money, and energy to develop offensive cyber-warfare capabilities, including the capacity to engage in state-sponsored denial-of-service attacks, and the use of Trojan horses, viruses and worms." I wish I had been more vociferous.

The anti-terrorism laws proclaimed after 9/11 were, in the main, defensive in nature and many had sunset clauses attached because they were considered extraordinary, extrajudicial measures in a time of existential crisis. However, especially vis-à-vis cyberspace, this defensive posture quickly morphed into developing offensive capabilities. With rare exception, those laws are still with us and have been enhanced. In this domain the only thing the sun appears ready to set on is the right to communicate and share information privately.

• • •

Founded in the spring of 2001, just prior to the incendiary events later that year, the Citizen Lab's mission was to combine technical interrogation, field research, and social science to lift the lid on the Internet. It remains so to this day. We aim to document and expose the exercise of power hidden from the average Internet user, and we do so basically by using the same practices as state intelligence agencies – by combining technical intelligence and field investigations with open-source information gathering. Our intent is to "watch the watchers" and to deliver our findings to the public, to constantly probe the degree to which cyberspace remains an open and secure commons for all. Situated at the University of Toronto, the Citizen Lab has the protection, resources, and credibility it needs to do what it must do, and from this base we have built international partnerships with researchers and universities around the world. We have eyes in many places and have become a digital early warning system, peering into the depths of cyberspace and scanning the horizon. What we have seen and continue to see is disturbing.

• • •

Another word, a few words actually, about the title.

In 1999, Stanford University's Lawrence Lessig published a book called *Code and Other Laws of Cyberspace*. Its central thesis is that the instructions encoded in the software that effectively run the Internet shape and constrain what is communicated just as laws and regulations do. Although Lessig did not emphasize it, that thesis is part of a larger tradition of theorizing about communications technology associated most prominently with Canadian academics Marshall McLuhan ("the medium is the message") and Harold Innis ("the bias of communications"). According to this tradition, communications technologies are rarely neutral and their material properties – the wires, cables, machines themselves, and so forth – have direct societal impacts. Think about this for a moment. To what degree does the machine in front of you, that you log onto and operate daily, now determine your behaviour, what you do and don't do? In many ways changes in modes of communication are like changes in ecological systems, with ideas, social forces, and institutions analogous to species. And when the ecology of communications changes, some species flourish and thrive, others wither and die.

Although Lessig uses the term "code" in a literal sense to refer to actual software, in this book I use it more metaphorically, to refer to the infrastructure of cyberspace, from the invisible spectrum of electromagnetic waves to the vast amounts of plastic, metal, and copper that now surround us, to the trillions of lines of spaghetti-like instructions – the actual codes – that keep it all functioning. Like Lessig, I believe that cyberspace is not an empty vessel or neutral channel. How it is structured matters for identity, human rights, security, and governance . . . and we need to tend to it to preserve it as a secure and open commons.

The word "black" conjures up that which is hidden, obscured from the view of the average Internet user. Never before have we been surrounded by so much technology upon which we depend, and never before have we also known so little about how that technology actually works. I am not talking about programming a VCR, or lifting the hood of your car in the faint hope that you can fix the engine, or trying to brew a cup of coffee from a digitally operated espresso machine. I am talking about an intimate and ongoing understanding of what's going on beneath the surface of the systems upon which we have become so reliant in order to communicate and remain informed.

The science fiction writer Arthur C. Clarke argued that "any sufficiently advanced technology is indistinguishable from magic," and as cyberspace grows more and more complex the more it becomes for most people a mysterious unknown that just "works," something we just take for granted. It is not only that we know less and less about the technical systems upon which we depend, the problem is deeper than that. We are actively discouraged, by law and the companies involved, from developing a curiosity about and knowledge of the inner workings of cyberspace. The extraordinary applications that we now use to communicate may feel like tools of liberation, but the devil is in the details, in the lengthy licence agreements that restrict how they can be used. And while exploring that technology is strictly policed, and sometimes carries with it warranty violations, fines, even incarceration, the spread of black-code-by-design is a recipe for the abuse of power and authority, and thus protecting rights and freedoms in cyberspace requires a reversal of that taboo, a spotlighting on that which is hidden beneath the surface.

"Black" also refers to the criminal forces that are increasingly insinuating themselves into cyberspace, gradually subverting it from the inside out. The Internet's original designers built a system of interconnection based on trust, and as beautiful as that original conception was, how it might be abused was never predicted, could not be predicted. One of the first Internet applications, email, was almost instantly hijacked by the persistent nuisance of spam. Each subsequent application has followed suit, and with the almost wholesale penetration of the Internet into homes, offices, governments, hospitals, and energy systems, the stakes are much higher, the consequences of those malignant forces much more serious. Those who take advantage of the Internet's vulnerabilities today are not just juvenile pranksters or frat house brats; they are organized criminal groups, armed militants, and nation states. Add to this mix the demographic shift that is occurring and the picture gets more frightening. Most of the world's future Internet population will live in fragile, and in many cases corrupt, states.

And then there are the secretive worlds of national defence and intelligence agencies, as in "black ops," "black budgets," going "deep black" – worlds that have now become major players in cyberspace security and governance. The collection of three-letter agencies born alongside World War II (CIA, FBI, NSA, KGB, etc.) that became global behemoths during the Cold War may have seemed to be on the edge of extinction in the 1990s, but the combination of "big data" (the massive explosion of digital information in all of its forms), security threats, and the spectre of terrorism has created a power vortex into which these agencies, with their unique information-gathering capabilities, have stepped.

At the very moment when we are surrounded with so much access to information and apparent transparency, we are delegating responsibility for the security and governance of cyberspace to some of the world's most secretive agencies. And just as we are entrusting so much information to third parties, we are also relaxing legal protections that restrict security agencies from accessing our private data, from investigating us. The title *Black Code* refers to the growing influence of national security agencies, and the expanding network of contractors and companies with whom they work.

• • •

The Internet began with the spirit of "hope springs eternal." Today, sadly, we live in a time of cyber phobia. Cyber espionage and warfare, the growing menaces of cyber crime and data breaches, and the rise of new social movements like WikiLeaks and Anonymous have vaulted cyber security to the top of the international political agenda, at untold cost. Almost every day a new headline screams about a serious problem in cyberspace that demands immediate attention. There is a palpable urgency to act – to do something, anything.

As ominous as the dark side of cyberspace may be, our collective reaction may become the darkest driving force of all. Fear is becoming the dominant factor behind a movement to shape, control, and possibly subvert cyberspace, and "What begins in fear usually ends in folly," as English poet Samuel Taylor Coleridge put it.

We stand at a precipice where the great leap in human communication and ingenuity that gave us global cyberspace could continue to bind us together or deteriorate into something malign. Only by fully uncovering the battle for the future of cyberspace can we understand what's at stake, and take steps to ensure that this degradation of one of humanity's greatest innovations does not happen.

(An interesting sidebar to this discussion ... "Mainstream media" are often criticized for only following horse races – elections, scandals, and so on – and for giving scant treatment to deep, difficult issues. Regarding cyberspace governance and security, I have actually found that mainstream outlets like the *New York Times*, *Bloomberg News*, *Wall Street Journal*, and others, have done, all things being equal, solid reporting and have been receptive to Citizen Lab investigations and reports. Even though the conceit in much of cyberspace is that media "organs of the establishment" are beholden to special interests and their advertisers, I have not found this to necessarily be the case. The more important matter is that if these issues are out there, reported on in the mainstream press, why are so few people paying attention?)

INTRODUCTION
Cyberspace: Free, Restricted, Unavoidable

Look around you. Do you see anyone peering into their smartphone? How many times have you checked your email today? Have you searched for a wifi café to do so? How many people have you texted? Maybe you're a contrarian, don't own a smartphone. You find all this "connectivity" to be a social menace that isolates people from the world around them, as they stare endlessly into the glow of their computer screens, or engage in loud conversations with invisible others as they walk down the street gesticulating. If your date answers that cellphone call all is lost, you think. The digital revolution is not all that it's cracked up to be, you say, and you resist it.

Good luck with that.

Even those of you who resist or fear cyberspace sense that we are in the midst of an onslaught. And we are! You resist initially because it is drawing you in, inevitably. Whether you like it or not, to remain part of civil society you have to deal with it. Cyberspace is everywhere. By the end of 2012 there were more mobile devices on the planet than people: cellphones, laptops, tablets, gaming consoles, even Internet-connected cars. Some estimates put the number of Internet-connected devices now at 10 billion. Cyberspace has become what researchers call a "totally immersive environment," a phenomenon that cannot be avoided or ignored, increasingly embedded in societies rich and poor, a communications arena that does not discriminate. Connectivity in Africa, for instance, grows at some 2,000 percent a year. While the digital divide remains deep, it's shrinking fast, and access to cyberspace is growing much faster than good governance over it. Indeed, in many regions rapid connectivity is taking place in a context of chronic underemployment, disease, malnutrition, environmental stress, and failed or failing states.

Cyberspace is now an unavoidable reality that wraps our planet in a complex information and communications skin. It shapes our actions and choices and relentlessly drives us all closer together, drives us even towards those whom, all things being equal, we would rather keep at a distance: A shared space, a global commons, the public square writ large. You've heard all the ecstatic metaphors used by enthusiasts and your thoughts turn elsewhere. "Hell is other people," Jean-Paul Sartre famously wrote in *No Exit*, and now teeming billions of them are potentially in your living room, or at least in your email inbox, that silent assassin. You cherish your privacy.

Of course, there have been previous revolutions in communications technology that have upset the order of things and caused outrage and celebration. The alphabet, the invention of writing, the development of the printing press, the telegraph, radio, and television come to mind. But one of the many things that distinguishes cyberspace is the speed by which it has spread (and continues to spread). Those other technological innovations no doubt changed societies but in an "immersive" sense only over many generations, and more locally than not. Cyberspace, on the other hand, has connected two-thirds of the world – has joined, that is, more than 4 billion people in a single communications environment – in less than twenty years. And it is moving onward, accelerating in fact, bringing legions into its fold each and every day.

The amount of digital information now doubles every year, and the "information superhighway" might be best described as continuous exponential growth, more on-ramps, more data, all the time, faster, more immediate, more accessible, its users always on, always connected. This speed and volume make getting a handle on the big picture difficult, and the truth is – a hideous truth, especially for those of you who think of yourselves as "off the grid," somehow away from the connected world, and proudly disconnected – is that no one is immune. Let's imagine for a moment that you don't own a computer, have never sent an email or text, and don't know what "app" means. The thing that informs you, that prepares you for cocktail parties and other gatherings, is mainstream or "old" media – newspapers, radio, and TV. Look closely at this "old media": How much of it is now "informed by," even directed by, "new media," by thousands, even millions, of "citizen journalists," unpaid, unaccountable, but with cellphone cameras permanently at the ready, documenting events as they happen in real time, unfiltered, and, perhaps, unreliable. The other truth is that no one really knows what this hurricane will leave behind or where it will take us. We're just struggling to hang on.

Another chief difference between then and now is that today, through cyberspace, it is us, the users, who create the information, do the connecting, and sustain and grow this unique communications and technological ecosystem. Save for the telephone, previous communications revolutions required a certain passivity on the part of consumers. There was little or no interactivity. We turned on the radio and listened, watched television happy to tune out and not to have to respond. The information provided, even the news of the day, simply washed over us. (We might get a call from a ratings agency, might be polled, might write a letter to the editor, but in the main we were passive recipients not active participants.) Cyberspace is wholly different, and potentially far more egalitarian. It is the lonely man in a café clicking away, the mother out for dinner with friends discreetly contacting her kids, the armed militant in Mogadishu, the criminal in Moscow, as much as it is any-one or any institution in particular, who feed the machine, cause it to grow, to envelop us further. While it is difficult to pin down a constantly moving target, this much can be said: it is peculiar to cyberspace that we, the users, shape it as much as we are shaped by it. We are at it every day, every night, transforming it all the while. Cyberspace is what we make of it. It is ours. We need to remember this before it slips through our grasp.

This remains the issue. One of the extraordinary – and for many liberating – things about cyberspace is that while massive and hugely profitable corporations like Apple and Google have made it possible and accessible (virtually) to all, they don't actually control it. Indeed, while having seeded the terrain, Apple, Google, and other gigantic corporations might have no greater control over cyberspace than those of us operating alone, at home, at our computer screens. This generative quality changes everything, causes grave concern, causes many to demand that cyberspace be brought under control.

• • •

It's difficult not to marvel at the extraordinary benefits of cyberspace. To be able to publish anything and have it immedi-ately reach a potential worldwide audience represents a democra-tization of communications that philosophers and science fiction writers have dreamed about for centuries. Families continents apart now share in each other's daily struggles and triumphs. Physicians connect with patients thousands of kilometres away, in real time. Through vast aggregations of data we can now predict when disease outbreaks are likely to occur, and take precautionary measures. We can pinpoint our exact longitude and latitude, iden-tify the nearest wifi hotspot, and notify a friend that we are, well, nearby and would like to meet.

But there is a dark side to all this connectivity: malicious threats that are growing from the inside out, a global disease with many symptoms that is buttressed by disparate and mutually reinforcing causes. Some of these forces are the unintended by-products of the digital universe into which we have thrust ourselves, mostly with blind acceptance. Others are more sinister, deliberate manipulations that exploit newly discovered vulnerabilities in cyberspace. Together they threaten to destroy the fragile ecosystem we have come to take for granted.

Social networking, cloud computing, and mobile forms of connectivity are convenient and fun, but they are also a dangerous brew. Data once stored on our actual desktops and in filing cabinets now evaporates into the "cloud," entrusted to third parties beyond our control. Few of us realize that data stored by Google, even data located on machines in foreign jurisdictions, are subject to the U.S. Patriot Act because Google is headquartered in the United States and the Act compels it to turn over data when asked to do so, no matter where it is stored. (For this reason, some European countries are debating laws that will ban public officials from using Google and/or other cloud computing services that could put their citizens' personal information at risk.) Mobile connectivity and social networking might give us instant awareness of each other's thoughts, habits, and activities, but in using them we have also entrusted an unprecedented amount of information about ourselves to private companies. We can now be tracked in time and space with a degree of precision that would make tyrants of days past envious – all by our own consent. Mobile devices are what Harvard's Jonathan Zittrain, author of *The Future of the Internet*, calls "tethered appliances": they corral us into walled gardens controlled by others, with unknown repercussions.

These technological changes are occurring alongside a major demographic shift in cyberspace. The Internet may have been born in the West but its future will almost certainly be decided elsewhere. North Americans and Europeans make up less than 25 percent of Internet users, and the West in general is almost at saturation point. Asia, on the other hand, comprises nearly 50 percent of the world's Internet population (the most by region), and only 28 percent of its people are online (next to last by region). Some of the fastest growth is happening among the world's weakest states, in zones of conflict where authoritarianism (or something close), mass youth unemployment, and organized crime prevail. How burgeoning populations in Africa, Asia, the Middle East, and Latin America will use and shape cyberspace is an open question.

The young "netizens" who launched the Arab Spring were born into a world of satellite broadcasts, mobile phones, and Internet cafés. They were plugged in to the digital world and able to exploit viral networks in ways difficult for authorities to anticipate or control. Meanwhile, perhaps the most innovative users of social networking and mobile technologies in Latin America today are the drug cartels, which use these tools to instill fear in citizens and lawmakers, intimidate journalists, and suppress free speech. To understand how and in what ways cyberspace will be used in the years to come we need to analyze innovation from the global South and East, from users in cities like Tegucigalpa, Nairobi, and Shanghai, the new centres of gravity for cyberspace.

• • •

And then there is cyber crime, a part of cyberspace since the origins of the Internet, but now explosive in terms of its growth and complexity. The economy of cyber crime has morphed from isolated acts by lone "basement" criminals into a highly professionalized transnational enterprise worth billions annually. Every day, security companies must review thousands of new samples of malicious software. Botnets that can be used for distributed denial-of-service (DDOS) attacks against any target can be rented from public forums and websites for less than $100. Some even offer 24/7 technical help. Freely available spyware used to infiltrate networks has now become commonplace, a mass commodity. As a result, the people who maintain network security for governments, banks, and other businesses face a continuous onslaught of cyber-crime attacks.

Cyberspace has evolved so quickly that organizations and individuals have yet to adopt proper security practices and policies. We have created a hyper-media environment characterized by constant innovation from the edges, extensive social sharing of data, and mobile networking from multiple platforms and locations, and in doing so, we have unintentionally opened ourselves up to multiple opportunities for criminal exploitation. Cyber crime thrives partly because of a lack of controls, because the criminals themselves can reap a digital harvest from across the globe and hide in jurisdictions with lax law enforcement and regulations. Furthermore, it moves at the speed of electrons, while international law enforcement moves at the speed of bureaucratic institutions. It is almost routine now to hear about cyber criminals living openly in places like St. Petersburg, Russia, and exalted as tech entrepreneurs, not the digital thugs that they are.

No doubt, cyber crime is a major nuisance, a shadowy, unregulated economy that costs decent folks dearly, but even more disturbing is how cyber crime, espionage, sabotage, and even warfare appear to be blurring together. Almost daily, there are breaches against government departments, private companies, or basic infrastructure. The Citizen Lab has investigated several of these cases, two of which we documented in our reports, *Tracking GhostNet* and *Shadows in the Cloud*. The victims, all compromised by China-based perpetrators, included major defence contractors, global media outlets, government agencies, ministries of foreign affairs, embassies, and international organizations like the United Nations.

How far down this road have we gone? A 2012 *New York Times* report revealed that the United States and Israel were responsible for the Stuxnet virus, which sabotaged Iranian nuclear enrichment facilities in June 2010. While the two countries remained mum about the charge, they did not deny it. The incident represents the first time governments have tacitly acknowledged responsibility for a cyber attack on the critical infrastructure of another country, a de facto act of war through cyberspace.

The techniques used in these state-based breaches and attacks are indistinguishable from those used by cyber criminals. Indeed, Stuxnet has been described as a "Frankenstein" of existing cyber-crime methods and tradecraft, and many now see cyber crime as a strategic vector for state-based and corporate espionage. Hidden in the shadows of low-level thuggery and cyber crime for cash, in other words, are more serious and potentially devastating operations, like acts of sabotage against critical infrastructure. Now perilously networked together, such infrastructure is especially vulnerable to cyber attacks: our smart grids, financial sectors, nuclear enrichment facilities, power plants, hospitals, and government agencies are all there for the taking. And this is happening at a time when militaries, criminal organizations, militants, and any individual with an axe to grind are refining capabilities to target and disrupt those networks. Cyberspace has become a battleground, a ground zero, for geopolitical contests and armed struggle.

Cyber crime is much more than a persistent nuisance. It has become a key risk factor for governments and businesses. The consequences of this exploding threat are numerous and wide-ranging and have led to greater and greater pressures for state regulation and intervention. Proliferating cyber crime and espionage have vaulted cyber security to the top of the international political agenda and brought about a sea change in the way that governments approach cyberspace. Where once the dominant descriptor of Internet regulation was "hands off," today the talk is all about control, the necessary assertion of state power, and, increasingly, geopolitical contestation over cyberspace itself.

The OpenNet Initiative (ONI), a project in which the Citizen Lab participates and that documents Internet content filtering worldwide, notes that roughly 1 billion Internet users live in countries (over forty of them) that regularly censor the Internet. States have become adept at content-control regulations, mostly downloading responsibilities to the private sector to police the Internet on their behalf, but some governments have gone further, engaging in offensive operations on their own, including disabling opposition websites through DDOS or other attacks, and/or using pro-government bloggers to flood (and sometimes disable) the information space.

Although conventional wisdom has long maintained that authoritarian regimes would wither in the face of the Internet (and some in the Middle East and North Africa appear to have done so), many have turned the domain to their advantage. Tunisia and Egypt may have succumbed to Facebook-enabled protestors, but China, Vietnam, Syria, Iran, Belarus, and others have successfully employed second- and third-generation control techniques to penetrate and immobilize opposition groups and cultivate a climate of fear and self-censorship. These states are winning cyberspace wars. For them "Internet freedom" is just another excuse for state control.

• • •

It would be wrong, however, to see the growing assertion of state power in cyberspace as coming only from authoritarian regimes. As Stuxnet suggests, cyberspace controls, in fact, are being driven and legitimized just as much by liberal democratic countries. Many liberal democratic governments have enacted or are proposing Internet content-filtering laws, mostly, they say, to clamp down on copyright infringements, online child pornography, or other content deemed objectionable, hateful, or likely to incite violence. Many have also pushed for new surveillance powers, downloading responsibilities for the collection of data onto the private sector while relaxing judicial oversight around the sharing of information with law enforcement and intelligence agencies. They are also developing offensive information operations. The United States and many other Western governments now speak openly about the need to fight (and win) wars in this domain.

Not surprisingly new companies have sprouted up to serve the growing pressure to "secure" cyberspace, a growth industry now worth tens of billions of dollars annually. Countries that censor the Internet have usually relied on products and services developed by Western manufacturers: Websense in Tunisia, Fortinet in Burma, SmartFilter in Saudi Arabia, Tunisia, Oman, and the United Arab Emirates. Filtering and surveillance devices manufactured by Blue Coat Systems, an American firm, have been found operating on public networks in Afghanistan, Bahrain, Burma, China, Egypt, India, Indonesia, Iraq, Kenya, Kuwait, Lebanon, Malaysia, Nigeria, Qatar, Russia, Saudi Arabia, Singapore, South Korea, Syria, Thailand, Turkey, and Venezuela – a list that includes some of the world's most notorious human rights abusers. Netsweeper, a Canadian company, sells censorship products and services to ISPs across the Middle East and North Africa, helping regimes there block access to human rights information, basic news, information about alternative lifestyles, and opinion critical of the regimes. In 2012, dissidents in the United Arab Emirates and Bahrain were shown, during interrogations where they were arrested and beaten, transcripts of their private chats and emails, their computers obviously compromised by their own government security agencies. Those agencies didn't use an off-the-shelf piece of cyber-crime spyware to do the job; rather, they employed a high-grade commercial network intrusion kit sold to them by British and Italian companies.

American, Canadian, and European firms that used to brag about connecting individuals and wiring the world are now turning those wires into secret weapons of war and repression. Suddenly, policy-makers are being given tools they never before imagined: advanced deep packet inspection, content filtering, social network mining, cellphone tracking, and computer network exploitation and attack capabilities.

This is not the way it was supposed to be.

As the imperatives to regulate, secure, and control cyberspace grow, we risk degrading (even destroying) what made cyberspace unique in the first place. In the face of urgent issues and real threats, policy-makers may be tempted to lower the bar for what is seen as acceptable practice or, worse, throw the baby out with the bath water. Before extreme solutions are adopted we must address the core value that underpins cyberspace itself: ensuring that it remains secure, but also open and dynamic, a communications system for citizens the world over.

1.

Chasing Shadows

"I'm in."

"What do you mean, you're in?"

"I've got full access to the control panel. There's a list of computers here that looks pretty serious. It's much more than just the Dalai Lama's office."

It started as an experiment, another wild hunch. We had been working with computer hackers and field researchers the world over for years, picking up the digital trails left by state officials and a slew of bad guys. But this was different. It was January 2009, and Nart Villeneuve, the then thirty-four-year-old lead technical researcher at the Citizen Lab, had made an extraordinary breakthrough. "I'm in," he whispered into the phone from his workstation, and when I asked how, he said, "I just Googled it."

So began the story of GhostNet. Villeneuve's finding – twenty-two characters typed into Google – turned out to be our Rosetta Stone, our key to eventually uncovering an espionage network affecting more than 100 countries and targeting ministries of foreign affairs, embassies, and other state agencies, international organizations, businesses, and global media outlets. My world would never be the same.

The GhostNet investigation had begun months earlier, when Greg Walton, one of our field researchers, learned of persistent concerns about computers being hacked into at the Dalai Lama's headquarters. Walton knew northern India well, had lived in the small town of Dharamsala, where the Tibetan Government-in-Exile, Tibetan NGOs, and the Office of His Holiness the Dalai Lama are located. The Tibetan community in exile had long suspected that their computers were being monitored by the Chinese government. While attempting to cross the border into China, people doing advocacy work on behalf of Tibet were detained, interrogated, and presented with transcripts of their private chat and email messages. Although it is possible – in fact, likely – that the Chinese government pressured companies to modify their products to provide them with backdoor access or to simply turn over user data upon request, it is also possible that the Tibetans had their computers compromised at source. Foreign government officials planning to visit the Dalai Lama, or to meet with him privately when he travelled to their countries, had been told by China to stand down, not to meet him. But the issue now was: how did Chinese authorities know *in advance* that this or that meeting between the Dalai Lama and foreign sympathizers was to take place?

When presented with the idea of the Citizen Lab checking into this matter, Tibetan officials agreed to turn over their machines for inspection. It was a serious decision, as we would be given unrestricted access to computers at the Office of His Holiness the Dalai Lama, the Tibetan Government-in-Exile, and Tibetan NGOs in Dharamsala, New York, Brussels, and London. Although the Dalai Lama himself liked to point out publicly that they "had no secrets," his office and those of other Tibetan organizations handled sensitive communications, including private correspondence and information about travel schedules. They took a risk working with us, one that paid off in the end.

• • •

Cyber espionage is a dark art, widely speculated about but rarely examined in the light of day. There have been cases of state cyber spying reported on in the media, but too often key pieces of evidence were either missing or, more likely, locked down in the secret chambers of the world's leading intelligence agencies. "Titan Rain," a huge compromise of American military and intelligence agencies and companies, was an exception between 2003 and 2006, and suspicions ran high that it was orchestrated by China-based hackers doing dirty work for their government. The Chinese government was almost certainly connected in some manner to what we unearthed too, and once the cat was out of the bag there would be international diplomatic furor.

While the Citizen Lab had been analyzing and exposing strange goings-on in cyberspace for years, the GhostNet investigation was unprecedented, the scope of the pilfering extraordinary. Computers based in the Dalai Lama's headquarters and Tibetan organizations were compromised, but so too were those in foreign government agencies, and in international organizations, companies, and media outlets the world over. Included among the victims were the ministries of foreign affairs in Iran, Bangladesh, Latvia, Indonesia, the Philippines, Brunei, Barbados, and Bhutan, and the embassies of India, South Korea, Indonesia, Romania, Cyprus, Malta, Thailand, Taiwan, Portugal, Germany, and Pakistan. Computers at the UN and ASEAN, and an unclassified computer located at NATO headquarters, were also attacked, as was the prime minister's office in Laos. One remarkable breach was of the mail server at the Associated Press office in Hong Kong, giving the GhostNet attackers access to emails sent to and from AP in Hong Kong containing information about stories before they were published.

For months we had a bird's-eye view of the attackers' command-and-control network, could see everything they were doing. They had made the mistake of not password-protecting all of their computer directories, assuming that no one would be able to access them if they were not linked to publicly. But Villeneuve spotted that string of twenty-two characters used repeatedly in the networking traffic collected from Tibetan organizations' computers, and on a hunch he copied then Googled it. Two results came up for obscure websites based in China, and he was then able to map almost all of the command-and-control infrastructure of the attackers, allowing us to see inside their operations without their knowledge. For weeks we watched transfixed, while an ever-expanding list of victims had their computers tapped, as cyber espionage on a massive scale unfolded in real time. We were able to isolate an individual at the Indian embassy in Washington, D.C., whose computer had been compromised by correlating data from the attacker's web interface with open-source information via Google, and this led us to his bio and contact information website. We thought about calling him with a warning – *unplug your computer now!* – but decided against doing so because we were concerned about tipping off the attackers. Better to analyze all of the data first, we thought. We were inside an international spy operation, the attackers and their hundreds of victims had no idea, and yet we were also, in our own way, engaging in a kind of cyber espionage.

We set up a sting operation by infecting an isolated computer at the Citizen Lab, our "honeypot," with the same trojan horse – a program in which malicious code is contained inside apparently harmless data – used by the attackers. Then we waited. A few days later our honeypot lit up. A visitor was poking around. He came and went quickly, but stayed just long enough for us to see that he was connecting from a digital subscriber line (DSL) through an IP address on Hainan Island, the same location as one of the command servers, which happened to be a government of Hainan computer. Hainan Island is home to the Lingshui signals intelligence facility and the Third Technical Department of the Government of

China's People's Liberation Army (PLA). Established in the 1960s, and upgraded substantially in the 1990s, the signals intelligence facility is staffed by thousands of analysts, and its primary mission is to monitor U.S. naval activity in the South China Sea. (It's a big island, to be sure, but that a signals intelligence facility of some renown happens to be located there is intriguing.)

The tool used to hack into government agencies, media outlets, and others, was a trojan called Ghost RAT that gave the attackers the ability to remove any file from the computers under their control. (RAT stands for "remote access trojan.") We had seen this through Greg Walton's monitoring of the network traffic of Tibetan organizations – connections were then made to China-based IP addresses, hidden from view, and sensitive documents were plucked right out from under the noses of unwitting computer users. Ghost RAT also gave the attackers the ability to record every keystroke entered into the infected computers, capture all passwords and encrypted communications, and turn on audio and video capture devices. Effectively, it could turn the machines under their control into wiretaps.

Remarkably, most of the GhostNet spying capabilities are freely available through an open-source network intrusion tool, the same Ghost RAT that anyone, to this day, can download from the Internet. With widely available and easy-to-access tools like Ghost RAT we have entered the age of do-it-yourself cyber espionage.

• • •

"Who done it?" The obvious answer was China. The geographic locations of most victims formed a crescent moon around China's southern flank and read like a who's who of its most important strategic adversaries: Tibetans, Russians, Iranians, Vietnamese, and so on. We had something of a smoking gun with the Hainan Island sting, but we needed to be sure, needed to articulate precisely how these types of attacks could be launched by just about anyone, and, perhaps most importantly, by people who might have an interest in making it *look as if* the Chinese government was responsible. Having gained access to the attackers' command-and-control interfaces would have allowed us, for instance, to infiltrate the same organizations, and no one would have been the wiser. We had a list of the compromised computers and knew where the vulnerabilities lay. It would have been easy for us to commandeer those computers, and there were many agencies that would pay for access to, say, the Iranian foreign affairs ministry or the Indian embassy in Washington. (Later, I would meet computer security engineers who had monetized that type of access and knowledge, selling information about specific target vulnerabilities to, presumably, law enforcement and intelligence agencies for a king's ransom.) Although the attacks emanated from China's Internet space they could have originated from a garage in New Jersey. In fact, one of the command servers was in the United States. In short, GhostNet could have been orchestrated and controlled by anyone, anywhere.

Cyber security has long been highly politicized and dozens of government agencies and transnational corporations have their irons in the fire, and are salivating at ever-increasing defence budgets for Internet surveillance. There is considerable vested self-interest in inflating the threat, and during our GhostNet probe (and ever since) our efforts have been to ensure accuracy and to establish a standard. Universities have a special role to play as stewards of evidence-based, impartial research on cyber security, and we needed to ensure that the GhostNet report weighed all of the available evidence as impartially as possible.

In the end, *Tracking GhostNet: Investigating a Cyber Espionage Network,* chronicled a landmark case in cyber espionage. The scope and importance of the victims, sophistication of the attack (given the negligible resources used to pull it off), detailed exposure of what was going on beneath the surface and, finally, the shock of such widespread infiltration made it so. We are used to our computers being windows onto the world. With GhostNet, we argued that "it is time to get used to them looking back at us."

· · ·

"It'll be on the front page," John Markoff of the *New York Times* told me hours before the GhostNet story appeared, and he was right. It was above the fold on Sunday, March 29, 2009, and soon thereafter became one of the top news stories in the world. The University of Toronto's media relations office was overwhelmed. There were satellite trucks parked outside of the Munk School of Global Affairs, where we are based, cameras everywhere, and I experienced my first media scrum. Later, I had to switch off my mobile phone because it never stopped ringing, and eventually I had to change my number altogether. While I was at the Citizen Lab, my home phone was barraged with calls; our children fielding messages in the early mornings from reporters in Europe and Asia just as confused as they were. There were surreal moments watching the Dalai Lama on television being asked to comment on our report, and Chinese government officials dismissing us as liars. Liu Weimin, the spokesman for the Chinese embassy in London, said the report was part of the Dalai Lama's "media and propaganda campaign," while foreign ministry spokesman Qin Gang said that we were haunted by a "Cold War ghost" and suffered from a "virus called the China threat."

"We have no secrets to hide," the Dalai Lama told CNN. "They should spy more, then they would know what we are doing." He soon got his wish. A few months later, our group (working this time with the U.S.-based volunteer computer security group, the Shadowserver Foundation) revisited the GhostNet campaign and returned to the Dalai Lama's headquarters to re-examine their computers. We found that they were thoroughly compromised, again, this time by a different China-based espionage campaign. We dubbed it the Shadow Network, "Shadows" for short. Although Shadows was largely restricted to India-related victims, this time we were able to recover copies of data stolen by the attackers as they were being removed from victims' computers. They had exfiltrated documents marked "Secret" from the Indian national security agency, private business information from Indian defence and intelligence contractors, and a year's worth of the Office of His Holiness the Dalai Lama's official and private correspondence with citizens, world leaders, and religious figures.

The GhostNet and Shadows probes (Shadows was also covered extensively in the media) exposed us to a subterranean world of political intrigue, but our findings were not entirely unexpected. We had been gathering evidence for nearly a decade, lifting the lid on the Internet and tracking a contest for the future of cyberspace that was becoming more intense with each passing year. The signposts were clear: cyberspace was changing fast, and not necessarily for the better.

2.

Filters and Chokepoints

"I have no idea what the Internet is!"

— Hayastan Shakarian, aged seventy-five

On March 28, 2011, the Internet went down in Georgia. For nearly twelve hours citizens had no access to Twitter, Facebook, their favourite YouTube videos, or their primary sources of news and online information. They could not access their online bank accounts or send emails. An information darkness had descended on the Eurasian country. The culprit? A nasty computer virus? Another Russian invasion? The latter would not be out of the question. Three years earlier, Georgia's Internet was brought to a halt as Russian ground troops invaded the territorial enclave of South Ossetia, the country's most contested region. Acting in support of the Motherland, scores of patriotic Russian hackers bombarded the Georgian Internet with a massive DDOS attack. It overwhelmed Georgian computers, including the government's websites and the country's banking and 911 systems.

As it turned out, the reason the Georgian Internet went dark this time around had to do with a seventy-five-year-old woman named Hayastan Shakarian, a "poor old woman" who had "no idea what the Internet is." She had been scavenging for firewood and old copper and accidentally cut a fibre-optic cable running parallel to a railway line, severing a key Internet connection. The effect was not limited to Georgia: because of how routing was configured in the region, Ms. Shakarian's inadvertent action also shut down the Internet in neighbouring countries. Ninety per-cent of Armenia's private and business Internet users were cut off, as were many in Azerbaijan.

• • •

What is cyberspace? Ask most people this question and they simply shrug: for them it remains a mysterious and technological unknown that "just works." The term *cyberspace* was coined in the early 1980s by science fiction writer William Gibson, who defined it as a "consensual hallucination," and that, indeed, is how it often seems. When we log onto Twitter or Facebook through our lap-tops or mobile phones, we enter into what feels like an ethereal world divorced from physical reality. Our thoughts about cyber-space – if indeed these can be characterized as thoughts at all – generally begin and end with the screen in front of us. We send an email and within seconds it magically appears on a friend's BlackBerry or laptop. We text a message and it is instantly received by a colleague on the other side of the world. We start up a video on YouTube and seconds later it is streaming in high definition. We take this for granted, don't even really think about it.

But what happens in those nanoseconds as the transmission of movies or emails or Internet searches are completed? Information travels at the speed of light, and the processing power of computers is astonishingly fast. It is almost impossible to grasp that the moment a text message is sent thousands of kilometres away the information is transmitted through a complex physical infrastruc-ture spanning multiple political jurisdictions, thousands of private companies and public entities, and numerous media of communi-cation, from wireless radio to fibre-optic cables, like the one Hayastan Shakarian accidentally severed in Georgia.

What if it were possible to overcome the laws of space and time and follow that email, text, or tweet? What would we see? Where does the data go? Who has access to it? What happens beneath the surface of cyberspace that we don't see? Although cyberspace may seem like virtual reality, it's not. Every device we use to connect to the Internet, every cable, machine, application, and point along the fibre-optic and wireless spectrum through which data passes is a possible filter or "chokepoint," a grey area that can be monitored and that can constrain what we can communicate, that can surveil and choke off the free flow of communication and information.

• • •

Those constraints begin the moment we interact with the Internet, starting with the instructions that make it all work. There are millions of software programs whose instructions shape and define the realm of the possible in cyberspace, and millions more are generated every year. Software, and its codes and commands, route traffic, run programs for us, let us into the virtual worlds we inhabit. One of the unique (and disconcerting for many) features of cyberspace is that anyone can produce software that can be distributed across the Internet as a whole. Some of the most ingenious pieces of code have been written by individuals for no other reason than to get their invention "out there," to boast and take advantage of a "free" distribution network.

Not all such code is benign. Countless thousands of ever-evolving malignant programs circulate through cyberspace as viruses, trojan horses, and worms. The implications of such "malware" range from minor inconveniences to threats to privacy to debilitating attacks on national security, and some researchers believe that there is now more malware than legitimate software applications, most of it emerging too quickly for computer security professionals to track. Malware ghettos inhabit vast and loosely connected ecosystems of insecure and outdated software programs, some of them lying dormant for years before being discovered. The progenitors prowl silently through social networking platforms, hijacking innocent people's Twitter or Facebook accounts to send phony requests to visit advertising sites or to do something more dastardly. Many of our computers may be infected by malware without our knowing it. What's worse, we pass these infections unwittingly along to friends and colleagues when we exchange information, visit malicious websites and blogs, or download documents from the Internet.

Much of the software that operates cyberspace is "closed" or proprietary, meaning that some person or company treats the code as its intellectual property. Open-source software, on the other hand, refers to code that is open to public inspection and sharing, depending on the licence. The tension between the two runs deep, and cuts across intellectual property and security issues. We may assume that closed code is relatively safe, but it is generally accepted by computer scientists that open-source code is more secure by virtue of having more "eyeballs" able to review it for potential flaws. An additional concern around closed code is the possibility that special instructions have been built in to it that might affect users without their knowledge – secret "backdoors" written into instructions by a defence or law enforcement agency, for example.

After software, the router – a device that sends information along to its destination – may be the second most fundamental chokepoint in cyberspace. Most of us are familiar with the small, often frustrating, boxes with tiny antennas that give us the ability to connect to the Internet wirelessly, whether in a coffee shop, or our homes and offices. In accessing these routers, we generally choose the default security presets provided by manufacturers without giving much thought to how easy they are to infiltrate. In a matter of minutes, armed with a $50 Alfa AWUS050NH USB wifi adapter (which can be purchased from Amazon) and a Linux security-testing application called BackTrack, a person without any computer engineering skills whatsoever could easily follow a set of simple instructions (laid out on YouTube, for instance) that would allow him or her to easily crack a Wireless Encryption Protocol (WEP)-enabled wifi router's password. Even simpler methods are available. Most wifi routers are shipped with default administrative passwords, accessible via a Web-based interface. Although users are cautioned to regularly change their passwords, most do not, allowing anyone to make intelligent guesses and access their routers remotely over the Internet. One website, called Router Passwords, archives known default passwords associated with router brands. How serious such vulnerabilities are was demonstrated in 2012 in Brazil, when attackers compromised 4.5 million home routers via default password hacks that changed people's DNS server settings so that when they attempted to visit websites like Google they were redirected to phony sites that looked legitimate but were in fact controlled by the attackers and contained malicious software.

Even without breaking into them, routers can leak information about us and our activities. In 2010, while mapping for its popular Street View service using its specially outfitted cars, Google collected information on wifi hotspots for use in a database it maintains to triangulate connections for mobile phones and other devices. It later emerged that Google had also collected (it claims unintentionally) payload information being secreted from unencrypted wifi routers along the way, including private information being communicated from homes and businesses. It turned out that its vehicles, outfitted with rooftop cameras and antennas, travelled up and down city streets like roving digital vacuum cleaners sucking up telephone numbers, URLs, passwords, emails, text messages, medical records, and video and audio files sent over open wifi networks.

In 2012, Cisco provided updates to its popular Linksys EA3500 dual-band wireless router. Users were redirected away from their usual administrative interface to "Cisco Connect Cloud" instead. In doing so, however, they had to agree to new terms of service that restricted use deemed "obscene, pornographic, or offensive," and that might "infringe another's rights, including but not limited to any intellectual property rights." (Cisco had also written in a clause that alluded to collecting all users' surfing history, but removed it after considerable outrage.) These limitations on what users can do in cyberspace were put in place not by their Internet service providers or by the government, but by the private manufacturer of the hardware they used to connect to the Internet.

Also in 2012, a cyber security researcher named Mark Wuergler found that Apple's iPhones transmitted to anyone within radio range the unique identifiers – known as MAC (media access control) addresses – of the last three accessed wifi routers. He cross-checked that information against a publicly accessible database of MAC addresses to pinpoint their locations on a map. Wuergler then created an application called Stalker to make it easier to harvest and analyze unintentionally leaked information – passwords, images, emails, and any other data transmitted by mobile phones and wifi routers. The information collected by Stalker contained the names of specific businesses regularly frequented, or friends and colleagues who are regular chat buddies. That information could be used to deceive someone into revealing further data which, in turn, could be used to undertake electronically based attacks.

The Citizen Lab uses a similar network analyzing tool, Wireshark, to sniff out hidden details of Internet traffic, though we do so only with the permission of those we monitor. Wireshark data has allowed us to see questionable connections being made to remote servers and evidence of malicious activity, as we found during our GhostNet probe. We have used the tool in workshops to demonstrate how much information can be gathered remotely without an Internet user's knowledge. Using Wireshark and connecting to a wifi network in a hotel, for example, one can collect information on who is attending a private meeting in a room down the hall (based on computer name data sent over the Internet) and sometimes usernames and passwords (if they are sent unencrypted). It is possible to collect data on all of the sites being visited and data downloaded by users in the room, the content of private chats, and updates to Twitter and Facebook accounts (again, if the user's communications are not encrypted).

We also use a tool called Nmap to scan networks and map the computers connected to them, which ports are open on those computers, what operating systems are used, et cetera. With Wireshark and Nmap employed together, we can precisely map the computers and devices logged onto a network (including all known vulnerabilities on those computers and devices), and collect much of what is being communicated by the people using those computers. All of this information can be collected – without the users ever noticing – by someone connecting to the same network a few metres down the hall using a few freely available open-source tools. Examples like these show how the multiplying access points into cyberspace can create unintentional vulnerabilities that may expose us to security and privacy risks.

• • •

We take cyberspace for granted. We assume that its basic modus operandi – uninterrupted connectivity to a shared communications environment – is always stable. That assumption is wrong.

Cyberspace is a highly dynamic ecosystem whose underlying contours are in constant flux. One of the most important recent changes has come about with the gradual movement away from searching the World Wide Web to a "push" environment where information is delivered to us instead, mostly through applications and services. A major impetus behind this shift has been the popularity of mobile devices, especially the Apple iPhone. Web browsers are functionally constrained by the smaller screens and other limitations of mobile devices, which has led to the popularity of applications that deliver specially tailored information to users instead. So, whereas in the past we might have visited the *New York Times* website via our browser, today a growing number of us download the *Times* app instead, signing off on another terms of service licence agreement in the process, and sharing with yet another third party a potentially far greater amount of personal data connected to our mobile phone. Of course, what can be "pushed out" can also be "pulled back" by companies, or turned off at the request of governments. Apple's iPhone, for instance, has a built-in remote "wipe" functionality that can permanently disable or erase the device and all of its apps.

When we communicate through cyberspace, our data is entrusted to the companies that own and operate the hardware, the applications and services, and the broad infrastructure through which our communications are transmitted and stored. These companies are the intermediaries of our Internet experiences, and what they do with our data can matter for how we experience cyberspace, and what we are permitted to do through it. They are critical agents in determining the rules of the road by virtue of the standards they insist upon, the operating decisions they take, and the constraints they impose on users. This is especially important as the volume of data they control becomes ever greater, ever more potentially lucrative in the global information economy.

The end-user licence agreements, terms of service, and other warranties we sign with these companies define what they can do with our data. Unfortunately, few users bother to read, let alone understand, them. It is hard not to be sympathetic. Unless one has an advanced legal degree, these documents are intimidating: tens of thousands of words in fine print, with exceptions and caveats that provide enormously wide latitude for what companies can do. Faced with this word-soup, most of us just click "I agree." What we are agreeing to might surprise us. Skype users, for instance, might be alarmed to find out that when they click on "I agree" to the terms of service they are assigning to Skype the right to change these terms at any time, at Skype's discretion, and without notice. Skype does not inform users about whether and under what conditions it will share user data with law enforcement or other government agencies. Users might not know that while they can stop using Skype, they cannot delete their accounts: Skype does not allow it.

The Internet is sometimes described as a massively decentralized and distributed "network of networks," a virtual place where information from everywhere is concentrated and accessible to all, an egalitarian thing of beauty. From one perspective, this description accurately characterizes its architecture. But within this network of networks there are critical chokepoints: a tangible, physical infrastructure that includes the hardware, software, cables, even the electromagnetic spectrum that exists in definable, real space. There are also regulatory and legal chokepoints: the ways in which cyberspace is structured by laws, rules, and standards that can facilitate forms of control. Mobile forms of connectivity, now the central method of communicating in cyberspace, are a case in point. The mobile industry is controlled by manufacturers of "closed" devices, handsets whose owners are prohibited from opening ("jailbreaking") or fiddling with their insides at the risk of warranty violations. Closed or proprietary software (with the exception of

Google's open-source Android operating system) means that the millions of lines of instructions that run a mobile phone's system are restricted to everyone but the company that sells the software. They operate through networks owned and serviced by a small number of ISPs and telecommunications companies (sometimes only one or two, depending on the region or country), and they function according to government-issued licences that set the conditions under which people can use the wireless spectrum. What from one perspective (that of the average user) looks like an ephemeral network of networks, from another (that of people in positions of authority) looks more like a tangible system of concrete controls through which power can be exercised and the nature of communications shaped for specific political or economic ends.

• • •

It is important to understand the political architecture of cyberspace because the companies that own and operate its infrastructure, applications, and devices are under increasing pressure from a variety of quarters to police the networks they manage: from the technical demands of managing increasingly complex types of communication flows like bandwidth-sucking video streams; from lucrative market opportunities to repackage and sell user data; from regulations passed down by governments to corporations to manage content and users. The latter are especially noteworthy because the Internet crosses political boundaries, and many companies have operations in multiple national jurisdictions, some of which do not respect the rule of law or basic human rights and whose policing of the Internet lacks transparency. To operate in some jurisdictions search engines, mobile carriers, and other Internet services are required to filter access to content deemed objectionable by host governments, turn services on and off in response to crises, push intimidating mass messages onto citizens living in certain regions, cities, or territories, and/or share information about users with state security services. More often than not the companies comply.

The cyberspace experience can vary dramatically depending on what application or device we use, which Internet café or hotspot we log on from, which ISP we contract with, and, most fundamentally, which political jurisdiction we connect from. Without the aid of special anti-censorship software, an Internet user in China is unable to connect to Twitter or Facebook, while a user in Pakistan cannot view YouTube. Users in Thailand cannot access videos on YouTube deemed insulting to the royal family. A user of the ISP du Telecom in the United Arab Emirates cannot access information about gay and lesbian lifestyles. (Using filtering technology produced by the Canadian company Netsweeper, such content is censored by du.) Indonesian users of BlackBerry devices are not able to access thousands of websites deemed pornographic and blocked by Research in Motion (RIM). Individuals living in volatile Kashmir are not able to access Facebook. According to ONI (OpenNet Initiative – a collaborative partnership of the Citizen Lab; the Berkman Center for Internet & Society at Harvard University; and the SecDev Group in Ottawa), dozens of governments now insist that ISPs operating in their political jurisdictions implement Internet censorship and surveillance on their behalf.

Internet filters and chokepoints can have bizarre collateral impacts on users' Internet experiences around "upstream filtering," cases where data transit agreements, or "peering," made between ISPs in separate countries can have spillover effects on Internet users in each others' countries. In 2012, ONI discovered that users in Oman were not able to access a large number of websites with Indian-related content (mostly Bollywood movies and Indian music). The source of the censorship, however, was not in Oman itself nor was it demanded by the government (for whom the sites in question were not controversial). Rather, it was the Indian ISP Bharti Airtel, with whom the Omani ISP, Omantel, has a peering arrangement.

This kind of collateral impact of Internet controls has a long history. In 2005, ONI found that when the Canadian ISP Telus blocked subscriber access to a website set up by a labour union intending to publicize its views about a dispute with Telus, it also unintentionally blocked access to over 750 unrelated websites. In 2008, the Pakistan Ministry of Information ordered Pakistan Telecom to block access to YouTube because of films uploaded to the site that purportedly insulted the Prophet Muhammad. In carrying out this order, Pakistan Telecom mistakenly communicated these routing instructions to the entire Internet, shutting down YouTube for most of the world for nearly two hours.

• • •

Most of the filtering described above takes place at the level of ISPs, the companies users contract with to get their basic connectivity. But there is a deeper layer of control, one that stretches down into the bowels of cyberspace: Internet Exchange Points (IXPs). While most users are familiar with ISPs, few have ever heard of IXPs. There are several hundred IXPs around the world: usually heavily guarded facilities with the level of security one encounters at an airport or defence installation. If you've ever wondered how it is that your email reaches your friend's email account with a completely different company, IXPs are the answer. It is here that traffic is passed between the networks of different companies – through border gateway protocols (BGP) exchanged between ISPs – and IXPs are the key strategic locations for the interception, monitoring, and control of large swathes of Internet communications. (In the early 2000s, I toured an IXP in downtown Toronto and saw row upon row of high-tech equipment, endless servers stacked on several floors. Down one long hallway there were hundreds of what appeared to be randomly distributed red tags attached to the equipment. I asked the tour guide, "What are the red tags?" He replied nonchalantly, "Oh, those are the wiretaps," and moved on.)

In 2002, Mark Klein, a twenty-year veteran technician with AT&T, was working at an IXP in San Francisco. He became suspicious after noticing some unusual activity in a "secure room" marked 641A. Klein was working in an adjacent area and had been instructed to connect fibre-optic cables to cables exiting from the secure room. He was not allowed to enter the room, and the people there were not the type of workers with whom Klein enjoyed lunch and coffee breaks. They kept to themselves and seemed to have special privileges. Later, Klein learned from his colleagues that similar operations were observed by engineers at other AT&T facilities across the United States.

Klein's suspicions eventually led to a class action lawsuit by the Electronic Frontier Foundation (EFF) against AT&T, alleging that the company had colluded with the National Security Agency (NSA) outside of the rule of law. As it turned out, inside room 641A was a data-mining operation involving a piece of equipment called Narus STA 6400, known to be used by the NSA to sift through large streams of data. The choice of location was significant. Because of the complex routing arrangements that govern the flow of traffic through cyberspace, many smaller ISPs sublease their traffic through AT&T – a globe-spanning "Tier 1" telecommunications company – and a large proportion of global communications traffic flows through its pipes. The AT&T-operated IXP in San Francisco is one of the world's most important chokepoints for Internet communications.

The IXP is a chokepoint for not only international traffic; it handles a large volume of domestic U.S. communications as well.

The NSA is prohibited from collecting communications from American citizens, and the data-mining operation at the AT&T facility strongly suggested that prohibition was being ignored. The EFF class action lawsuit took AT&T and another IXP operator, Verizon, to task for their complicity with what turned out to be a presidential directive instructing the NSA to install the equipment at key IXPs in order to monitor the communications of American citizens. In 2008, as the lawsuit dragged on, the Bush administration took pre-emptive action by introducing a controversial amendment to the Foreign Intelligence Services Act (FISA), giving telecommunications companies retroactive immunity from prosecution if the attorney general certified that surveillance did not occur, was legal, or was authorized by the president. This certification was filed in September of 2008 and shortly thereafter, the EFF's case was dismissed by a federal judge citing the immunity amendment. (Presidential candidate Barack Obama surprised many of his supporters by backing the FISA Amendment Act, and his administration has vigorously blocked court challenges against it ever since.) Although the full scope of the NSA's warrantless wiretapping program (code-named "Stellar Wind") is classified, William Binney, a former NSA employee who left the agency in protest, estimates that up to 1.5 billion phone calls, as well as voluminous flows of email and other electronic data, are processed every day by the eavesdropping system stumbled upon by Klein.

IXmaps, a research project at the University of Toronto, raises awareness about the surveillance risks of IXPs, particularly for Canadians. The project uses trace-routing technology to determine the routes discrete bits of information (or "packets") take to reach their destination over the Internet. In one example, IXmaps detailed the route of an email destined for the Hockey Hall of Fame in downtown Toronto and originating at the University of Toronto a few miles away. The email crossed into the United

States, was peered at an IXP in Chicago, and was probably exposed to one of the NSA's warrantless surveillance systems rumoured to be located at the facility. Known as boomerang traffic, this type of cross-border routing is a function of the fact that there are eighty-five IXPs in the U.S., but only five in Canada. Routing arrangements made by Canadian ISPs and telecommunications companies will routinely pass traffic into the U.S. and back into Canada to save on peering costs, subjecting otherwise internal Canadian communications to extraterritorial monitoring.

• • •

One of the long-standing myths about cyberspace is that it is highly resilient to disruption. For those of us who have laboured over Internet downtimes, email failures, or laptop crashes, this may seem like a fanciful idea. But the resiliency of cyberspace does have some basis in the original design principles of the Internet, whose architecture was constructed to route information along the most efficient available path and to avoid disruption in the event of a natural disaster (or nuclear attack). This resiliency was demonstrated in the aftermath of Hurricane Sandy in October 2012, which devastated the U.S. eastern seaboard and caused mass power outages, including the loss of local Internet and cell-phone connectivity. The network-monitoring company Renesys showed that the storm had collateral impacts on traffic as far away as Chile, Sweden, and India – but mostly in a positive sense: traffic destined for New York City that would have failed as a consequence of the storm was manually rerouted along alternative paths by savvy network engineers.

However, there are also many characteristics of cyberspace that demonstrate fragility and a lack of resiliency; Hayastan Shakarian's mistaken severing of an underground cable in Georgia to name one. It may come as a surprise that the same type of cables that Shakarian accidentally unearthed traverse the world's lakes and oceans, and bind cyberspace together in a very material sense. Undersea cables are one of the links that connect today's cyberspace to the late Industrial Revolution. The first such cables were laid in the late nineteenth century to facilitate telegraph traffic over long distances. Early designs were prone to failure and barely allowed the clicks of a telegraph exchange to be discerned across small bodies of water like the English Channel, but over time innovations in electronics and protective cable sheathings allowed the undersea cable industry to flourish. (This growth led to a dramatic increase in international telephone calls, and a new market for the sap of gutta-percha trees, which was used to coat and protect the cables until the mid-twentieth century.) Although international telecommunications have been supplemented with microwave and satellite transmissions, a surprisingly large volume of data still traverses the world through cables crossing the Atlantic and Pacific oceans, and major bodies of water like the Mediterranean Sea.

Due to the staggering costs involved, companies often share the same undersea cable trenches and sometimes competing companies even share the same protective sheathing. This makes those trenches highly vulnerable to major disruption. In a May 2012 article published on the website Gizmodo, provocatively titled "How to Destroy the Internet," the author details the physical elements of the Internet that could be easily targeted. He provides a link to a document alphabetically listing every single cable in the world, and its landing stations. While there are hundreds of cables, the total is not astronomical – and probably a lot fewer than what most people might expect for a network as vast as the global Internet. Among them is ACS Alaska-Oregon Network (AKORN), with its landing points in Anchorage, Homer, and Nikiski, Alaska, and Florence, Oregon; the Gulf Bridge International Cable System, with its landing points in Qatar, Iraq, Bahrain, Saudi Arabia, Oman, Iran, the United Arab Emirates, Kuwait, and India; and at the end of the long list, Yellow/Atlantic Crossing-2 (AC-2), which connects New York City to Bude in Cornwall, U.K. The author goes on to explain how many of the cables' onshore landing stations are sometimes "lying out on the sand like an abandoned boogie board," and how the cables could be severed with a few swings of an axe. Severing cables in this way at landing stations in only a few select locations – Singapore, Egypt, Tokyo, Hong Kong, South Florida, Marseilles, Mumbai, and others – could wreak havoc on most of the world's Internet traffic.

The 2006 Hengchun earthquake, off the coast of Taiwan, affected Internet access throughout Asia, and in 2008 two major cable systems were severed in the Mediterranean Sea. The cause of the severed cables is unknown, but some experts speculated that the dragging of a ship's anchor did the job. But a review of video surveillance taken of the harbour during the outage period showed no ship traffic in the area of the severed cable. Others suggested it could have been a minor earthquake, causing a shift in the ocean floor, but seismic data didn't support this conjecture. Whatever the cause, such cuts to cables are fairly routine: Even in their trenches, undersea cables are pushed to and fro by currents and constantly rub against a rough seafloor. In the case of the 2008 Mediterranean incident, the damage was severe: there were disruptions to 70 percent of Internet traffic in Egypt and 60 percent in India, and outages in Afghanistan, Bahrain, Bangladesh, Kuwait, the Maldives, Pakistan, Qatar, Saudi Arabia, and the United Arab Emirates. Nearly 2 million users were left without Internet access in the U.A.E. alone. Connections were not restored until a French submarine located the severed cables and brought them to the surface for repair.

Prior to the introduction of fibre optics, undersea cables were occasionally wiretapped by attaching instruments that collect radio frequency emitted outside the cables. During the Cold War, both the United States and the Soviet Union built special-purpose submarines that would descend on cables deep in the ocean and attach inductive coils to collect emissions. In his book *Body of Secrets*, historian James Bamford describes in detail Operation Ivy Bells in the early 1970s, in which the NSA deployed submarines in the Sea of Okhotsk to tap a cable connecting the Soviet Pacific Naval Fleet base in Petropavlovsk to its headquarters in Vladivostok. Specially trained divers from the USS *Halibut* left the submarine in frigid waters at a depth of 120 metres and wrapped tapping coil around the undersea cables at signal repeater points, where the emissions would be strongest. Tapes containing the recordings were delivered to NSA headquarters, and were found by analysts to contain extraordinarily valuable information on the Soviet Pacific Fleet. Several other submarines were later built for such missions, and deployed around the Soviet Union's littoral coastline and next to important military bases. When fibre-optic technology (which does not emit radio frequencies outside of the cable) was gradually introduced, the utility of such risky operations diminished. However, some intelligence observers speculate that U.S. and other signals intelligence agencies have capabilities to tap undersea fibre-optic cables by cutting into them and collecting information through specifically designed splitters.

• • •

Like undersea cables, satellites illustrate the fragile nature of cyberspace. In 2009, a defunct and wayward Russian satellite collided with an Iridium low Earth orbit satellite at a speed of over 40,000 kilometres per hour. The collision caused a massive cloud of space debris that still presents a major hazard. NASA's Earth observation unit tracks as many as 8,000 space debris objects of ten centimetres or more that pose risks to operational satellites. (There are many smaller objects that present a hazard as well.) The Kessler Syndrome, put forward by NASA scientist Donald Kessler in 1976, theorizes that there will come a time when such debris clouds will make near-Earth orbital space unusable. Although undersea fibre-optic cables provide the bulk of transit for global communications, they cannot sustain the entire load. A scenario such as the Kessler Syndrome, were it to come true, would end global cyberspace as we know it. Scientists have very few realistic solutions for cleaning up space debris.

Space is also an arena within which state intelligence agencies exercise power over the Internet. Although the Apollo missions were publicly justified on the basis of advancing human curiosity and science, the first missions into space actually had specific military and intelligence purposes. Since the 1960s, the superpowers have been developing globe-spanning satellites that are used for optical, infrared, thermal, and radar reconnaissance purposes. The Americans built a fleet of specially designed satellites whose purpose is to collect signals intelligence (sigint). Some sigint satellites operate in geostationary orbit 36,000 kilometres from the Earth's surface, and are used to zero in on radio frequencies of everything from microwave telephone signals to pagers and walkie-talkies. Such geostationary sigint satellites deploy huge parabolic antennas that are unfolded in space once the satellite is in position, with the signals being sent to NSA listening stations located in allied countries like Australia (Pine Gap), and Germany (Bad Aibling). Because the satellites operate in deep space, and radio signals travel in a straight line, radio frequencies can be collected efficiently and with little degradation. (Other sigint satellites take unusual orbits and can reportedly hover over regions of interest for longer periods and at lower altitudes.)

The NSA also operates sigint collection facilities at ground stations whose mission is to collect transmissions from civilian communications satellites. Typically, these enormous interception terminals, which look like giant angled birdbaths, are located in secure areas proximate enough to terrestrial transmission points to function properly. For example, one of the key signals intelligence stations in Canada is at the Canadian Forces Station Leitrim, just south of Ottawa, strategically positioned to intercept diplomatic communications moving in and out of the nation's capital.

Signals intelligence gathering is highly secretive, but it is a world we should all get to know better. Originally, the objects of sigint operations were other states' military and intelligence agencies: ballistic missile-test telemetry or operational instructions sent by high-ranking Politburo members. As the Cold War came to a close, however, this bipolar conflict atomized into a multitude of national security threats, some of which emanate from transnational terrorist groups and organized crime, and the scope of sigint operations became much broader and more widely dispersed across global civil society. As the volume of data flowing through global networks is exploding in all directions, and the tools to undertake signals intelligence have become more refined, cheaper, and easier to use, the application to cyberspace is obvious.

• • •

Although cyberspace is often experienced as an ethereal world separate from physical reality, it is supported by a very real infrastructure, a tangible network of code, applications, wires, and radio waves. Behind every tweet, chat message, or Facebook update, there is also a complex labyrinth of machinery, cables and pipes buried in trenches deep beneath the ocean, and thousands of orbiting satellites, some the size of school buses. In addition to being complex and fragile, this physical infrastructure contains a growing number of filters and chokepoints. Pulling back its layers is like pulling back curtains into dark hallways and hidden recesses, which, it turns out, are also objects of intense political contests.

There is another component of cyberspace, separate from its physical infrastructure, but that is also growing in leaps and bounds and becoming a critical part of the domain: the data. Information related to each and every one of us (and everything we do) is taking on a life of its own. It, too, has become an object of geopolitical struggle. Every call we make, every text and email we send, increasingly everything we do as we go about our daily lives, is recorded as a data point, a piece of information in the ever-expanding world of "Big Data" that is insinuating itself deeper and deeper into our lives and the communications environment in which we live.

3.

Big Data: They Reap What We Sow

From August 31, 2009 to February 28, 2010, German citizen Malte Spitz had virtually every moment of his life tracked — every step he took, where he slept and shopped, flights and train trips he booked, every person he communicated with, every Internet connection he made. All of his movements and communications were cross-checked against open-source information that could be found out about him, including his Twitter, blog, and website entries. The surveillance net around him was total, and all this information was dutifully archived. In short, someone, somewhere, knew Malte Spitz better than he knew himself.

Who was behind it? Was it the Bundesamt für Verfassungsschutz, Germany's formidable domestic intelligence agency responsible for monitoring threats to the German state? Did they plant a bug on him? Tap his phone lines? What did Spitz do to warrant such attention? Was he a criminal? A terrorist? A long-lost member of the 1970s-era Baader-Meinhof gang? None of the above.

Malte Spitz is a Green Party politician with a clean record. Deutsche Telekom, Germany's largest cellphone company, collected the data on him through his mobile phone, but it was Spitz himself (along with Germany's leading newspaper, *Die Zeit*) who collated it on an interactive map. He did so to demonstrate to the public the volume of data mobile carriers routinely collect about their users. Spitz asked Deutsche Telekom to send him all of the information they had on him. After several persistent appeals and the threat of a lawsuit, the company finally complied, sending Spitz a CD containing 35,830 lines of data. "Seen individually, the pieces of data are mostly inconsequential and harmless," wrote *Die Zeit*, "[but] taken together, they provide what investigators call a profile – a clear picture of a person's habits and preferences, and indeed, of his or her life."

• • •

On a daily basis, most of us experience a dynamic and interactive communications ecosystem that only two decades ago was the stuff of science fiction. And today, after perhaps a decade of near total immersion, it is almost impossible for most people in the West to imagine going back to a world before instant access and 24/7 connectivity, a bustling tableau of images, text, and sounds always at our fingertips. As with any such wholesale social change, we should expect unintended consequences, not all of them desirable. Past experiences with the printing press, telegraph, radio, and television tell us that new media environments shape and constrain the realm of the possible, favouring some social forces and ideas over others. The world of "big data" is no exception.

Understood by computer engineers as data sets that grow so large that they become awkward to work with and/or analyze using standard database management tools, I like to think of big data in metaphorical terms: as endless digital grains of sand on an ever-expanding beach that we produce as we act in cyberspace. Big data comes from everywhere: from space satellites used to gather climate information to lunchtime jokes on social media sites; from digital pictures and videos posted online to transaction records from grocery stores; from signals emitted by our mobile phones to information buried in the packet headers of our emails. Every day, 2.5

quintillion bytes of data are created, and 90 percent of the data in the world today was created in the past two years. According to Dave Turek, IBM's VP of exascale computing, from the beginning of recorded time up until 2003, humans created five "exabytes" of information (an exabyte = 1,000,000,000 gigabytes). In 2011, Turek estimates we produced that same amount of information every two days. IBM predicts that in 2013, we will be producing five exabytes every ten minutes. And it only grows. To take just one example: the Square Kilometer Array (SKA) telescope complex, currently under development by a consortium of countries and set to deploy in Australia in 2024, will produce one exabyte of data every day, roughly twice the volume of daily global Internet traffic in 2012.

Most of this data — searches, software downloads, music purchases, tweets, Skype calls, et cetera — comes from ordinary people going about their ordinary lives. In 2011, 200 million tweets were posted every day (and over 30 billion have been written and sent since Twitter's launch in 2006). Every sixty seconds, 168 million emails were sent, nearly 700,000 Google searches and Facebook status updates made, 375,000 Skype calls initiated, and 13,000 new iPhone apps downloaded.

Mobile forms of connectivity, including smartphones and tablets, have massively increased this volume of data. Being untethered to a fixed location allows us to be always on, always connected, always communicating. According to the multinational telecommunications company Cisco Systems, in 2012, and for the fifth year in a row, mobile data traffic more than doubled. It is expected that the number of mobile-connected devices will exceed the world's population by 2013; that is, there will be at least one operative mobile device for every human on the planet, and people will be constantly searching, texting, linking, networking, sharing, photographing, recording, purchasing. The proliferation of high-end handsets, tablets, and laptops on mobile networks, all major generators of data owing to the more detailed information experiences they support, will make up a greater proportion of the market. Says Cisco, a "single smartphone can generate as much traffic as 35 basic-feature phones; a tablet as much traffic as 121 basic-feature phones; and a single laptop can generate as much traffic as 498 basic-feature phones." Mobile data traffic is likely to grow at a compound annual rate of nearly 80 percent, reaching some eleven exabytes per month by 2016. We are immersed in a weightless but dense cloud of bits and bytes, percolating everywhere.

• • •

The data we create contains not just the information we send or interact with, but data about the data, or metadata. We rarely experience metadata directly, as it is buried in instructions and communications several layers below our interactions with our devices. But we can see it when we download photographs onto our computers or upload them to a photo-sharing site like Flickr. When we do so, we might notice that embedded in that digital photograph is data on the model of camera used to take the picture, the exact time it was taken, and the longitude and latitude of where it was taken (should such settings be activated by the user). Digital music and movie files typically contain metadata on the artist, album, date of the recording, and copyright information. Metadata on a mobile phone can contain information about a user's number, receiver's number, geographic coordinates of where the message was sent, date and time of the message, duration of any particular call, amount of data transmitted, and the cost of the transmission. Average users may have thousands of data points like these collected from them every day as they communicate through cyberspace. A typical smartphone emits a signal every few seconds, a "beacon" to nearby cellphone towers or wifi hotspots in order to triangulate the most efficient connection for the device. (It was this automatic beaconing that led Mark Wuergler to identify the security vulnerabilities on Apple products described in the previous chapter). Every call, text, or email we send via mobile phones yields space-time coordinates accurate to within metres of where we are and with whom we communicate. This information is stored on a server somewhere, or in multiple places – on "clouds" of computers – spread out across the physical infrastructure of cyberspace. It is an embodiment of us, a kind of cyberspace biography and activity chart, and we have little control over it.

Enthusiasts say that this world of big data is a gift. Google engineers, for instance, show us through their Flu Trends project how they can harness the information collected from millions of real-time queries to predict the location and timing of disease outbreaks. Simply by collating the number, location, and frequency of search queries for symptoms, insight of planetary significance and proportions is gained. If enough users in Chicago or San Francisco simultaneously search for information about a fever, Google can spot a virus before it spreads with a greater degree of accuracy than tools specifically designed to issue early-warning alerts employed by the U.S. Centers for Disease Control.

This ability to identify large-scale patterns can lead to new opportunities for humanitarian aid and development assistance, even in the most impoverished and dangerous of environments. In Haiti, for example, researchers used mobile-data patterns to monitor the movement of refugees and health risks following the massive hurricanes that slammed into that small island country in 2010. Crowd-sourcing data through the Ushahidi platform – a free and open-source software tool developed for information collection, visualization, and interactive mapping after the 2008 Kenyan election – is used to monitor elections, conflicts, and numerous other issues around the world. The LRA Crisis Tracker uses crowd-sourced data plotted on Ushahidi from radios distributed to local communities and other means to monitor atrocities undertaken by the Lord's Resistance Army (LRA), responsible for one of the most ruthless insurgencies in Africa. Each LRA-related incident is plotted on a map by type – civilian death, injury, abduction, looting – and once consolidated, the map shows the movements of the LRA across the region, and the scope, scale, and frequency of its actions. Incidents captured by cellphone cameras are linked to specific events on the website as corresponding evidence.

In Kibera, Nairobi, Kenya's largest slum, an experiment in crowd-sourcing data may revolutionize access to basic health care and sanitary services. Conditions in Kibera are dire: most residents are illegal squatters, and local officials regularly withhold basic services, including electricity, sewage treatment, and garbage collection. The most important commodity, water, is extremely scarce – turned on and off by capricious officials, and grossly overpriced by private dealers. Despite the poverty, over 70 percent of Kiberans have mobile phones. They are cheap, plentiful, and can save lives. Researchers at Stanford University are testing an app called M-Maji ("mobile water" in Swahili), which sends users text messages with up-to-date information on the location, price, and quality of water available from different vendors. They believe that this project can be replicated in impoverished communities around the world.

There are countless examples of big data being used to achieve such social goods, but such a rapid transformation of a global communications environment rarely avoids unanticipated negative consequences. To understand these, we need first to understand the political economy of big data, and this boils down to a simple question: Why are we able to use Gmail, Facebook, Twitter, and other cyber services for free?

• • •

"There is no free lunch," the old saying goes, and to that we should add "and no free tweet, either." The business model of big data rests on the repurposing of that which all of us routinely give away. Not surprisingly, the market to harvest the digital grains of sand on that constantly expanding beach has exploded: companies of all shapes and sizes systematically pick through our digital droppings, collating them, passing them around, inspecting them, and feeding them back to us. And this market shows no sign of slowing. In 2012, the open-source analyst firm Wikibon reported that the big-data market stood at just over $5 billion and predicted that it will grow to $50 billion by 2017. ISPs, web-hosting companies, cloud and mobile providers, massive telecommunications and financial companies, and a host of other new digital market organisms digest and process unimaginably large volumes of information about each and every one of us, each and every day, and it is then sold back to us as "value-added" products, services, or advertisements for yet more products and services!

Social networks may seem like secure, even cozy, playgrounds, but they are more like vacuum cleaners that hoover up every click and shared link, every status change, every tag and piece of personal history. As Facebook states frankly in its data-use policy, the company uses "the information we receive to deliver ads and to make them more relevant to you. This includes all of the things you share and do on Facebook, such as the Pages you like or key words from your stories, and the things we infer from your use of Facebook." Facebook "likes" are translated into customized dating and vacation ads; geolocation data is used to advertise local products. Not a single bit or byte is ignored: the companies involved reap what we sow. Freedom in cyberspace is just another word for nothing left unused.

Many network service companies stress the protections they put in place around customers' data. They insist that what is "theirs" is "yours" and use "I" and "my" as descriptors of their products and services. In practice, however, they treat our data as proprietary business records that they can retain, manipulate, and repurpose indefinitely. They see our habits (and us) as resources in the same way energy companies see untapped reserves of oil, for one simple reason: the online advertising industry is worth $30 billion annually. Whenever we surf the Internet today, depending on the browser we use and the settings we put in place on that browser, we give away pieces of ourselves. A tracking-awareness project, Collusion, has developed a plug-in for browsers that demonstrates how often such "sharing" takes place, usually without our knowledge. If I were to visit, say, http://www.washington-post.com, the Collusion plug-in shows that it shares information about my visit with twenty-one other websites. One of those sites is Scorecardresearch.com, and it sells beacons to participating websites (like washingtonpost.com), which place a cookie in visitor browsers. Cookies are small bits of text deposited on your browser that act as "unique identifiers" or signatures that give website owners details about visitors to their sites: their browsing histories, locations (based on IP addresses), and so on.

In 2012, the *Wall Street Journal* conducted a study of one of the "fastest-growing businesses on the Internet" – spying on Internet users. In their look at surveillance technologies that companies use to track consumers, they highlighted fifty of the most popular websites in the U.S., analyzed all the tracking files and programs these websites downloaded onto their test computers, and found that on average each website installed sixty-four tracking files, generally without warning. The website that downloaded the most tracking software was http://www.dictionary.com: 234 files onto the *Wall Street Journal*'s test computer. A Dictionary.com spokesperson said, "Whether it's one or ten cookies, it doesn't have any impact on the customer experience, and we disclose that we do it.

So what's the beef?" Users concerned about leaving digital traces of themselves all over the Internet might disagree.

The small print included with many applications and/or service contracts provides a window into the underlying reality of this market. By agreeing to terms and conditions contained in documents that scroll by on the way to the "I agree" button, users give the companies involved nearly unlimited permission to handle their data. In many cases involving mobile apps, users even give the developers the right to collect whatever images a camera happens to be focusing on, the image itself, as well as the phone's location. For example, the Facebook app developed by the Google Android smartphone, which has been downloaded more than 100 million times, has written into its terms of service the right for Facebook "to read SMS messages stored on your device or SIM card." The Flickr app can access location data, text messages, contact books, online account IDs, who a person is calling, and even the device's camera. In fact, the Flickr, Facebook, Badoo, Yahoo! Messenger, My Fitness Pal, and My Remote Lock apps can all access a user's entire contacts book and record who that user is calling. To repeat, the reason behind this data collection is advertising. As Daniel Rosenfield, director of the app company Sun Products testified in 2012: "The revenue you get from selling your apps doesn't touch the revenue you get from giving your apps away for free and just loading them with advertisements."

• • •

Few users realize how quickly big data about their communications accumulates in the hands of third-party operators. Malte Spitz is an exception. Max Schrems, an Austrian law student, is another. In 2011, Schrems asked Facebook to send him all of the data the company had stored on him. As he is European and Facebook's

European headquarters is in Dublin, Ireland, Schrems had the right to make such a request. Facebook dutifully sent him a CD containing 1,222 individual PDFs they had collected about him. The company had stored information on all of his logins, "pokes," chat messages, and postings, even those he had deleted. On a detailed map, it had also stored the precise geographical coordinates for all the holiday pictures (in which Schrems was tagged) that a friend of Schrems had taken and posted using her iPhone.

Schrems discovered that Facebook stores dozens of categories of data about its users so that it can accurately commodify its customers' digital persona for targeted advertisements. Some examples: the exact latitude, longitude, and altitude of every check-in to Facebook, which is given a unique ID number and a time stamp; every Facebook event to which a user has been invited, including all invitations ignored or rejected; and data on the machines used to connect to Facebook, so that Facebook can connect individuals to the hardware and software they use. Schrems eventually formed an activist group, Europe vs. Facebook, to launch complaints. This led to an inquiry by Irish privacy regulators and widespread media attention about the company's privacy policies. The battles continue.

This relentless drive for personal information leads to extraordinary encroachments on privacy by social networking companies and ISPs. Over the years, Facebook's default privacy settings have been continuously adjusted downwards, mostly in increments but sometimes dramatically. In 2005, only you and your friends could see your contact information and other profile data. Only your personal networks could see your wall posts and photographs, and nothing about you was shared by Facebook through the Internet. In 2007, an adjustment was made such that your personal network could see more of your profile data. And then, in early 2009, a major shift took place: suddenly, all Facebook users were permitted to see all of your friends, and the entire Internet could see your gender, name, networks, and profile picture. Another dramatic change took place in December 2009: Facebook settings were modified such that users' "likes" went from something exclusively seen by friends and friends-of-friends to the entire Internet. Months later, the same "all of the Internet" was extended to users' photos, wall posts, and friends. Like a giant python that has consumed a rat, Facebook captures, swallows, and slowly digests its users.

The search for new sources of personal information has led down other frightening paths. In 2010, the Sleep Cycle app was thrown onto the market. It monitors the sleep patterns of users from their mobile phones, and works when the phone is placed on the bed of the user. The app monitors movements and other patterns that determine periods of deep sleep, dreaming, and light sleep. Thirty minutes before the alarm is set to go off, it begins monitoring for the lightest periods of the sleep cycle and then gently nudges users awake with soothing sounds, instead of honking alarm bells. Data about the night's sleep is recorded and stored on the app's servers. (Naturally, the app also has an option to "share on Facebook.") Perhaps our dreams will be next, and then, worse, our nightmares.

The desire for big data is relentless, the temptations irresistibly strong, and in their lust for information about us many companies have disregarded basic privacy protections. Path, a popular social network, was caught uploading members' mobile phone contacts to its servers without permission. Twitter has admitted that it copied lists of email addresses and phone numbers from people who used its smartphone application. (And, again, the information was stored on its servers without users' permission.) A 2012 study by the mobile security company Lookout found that 11 percent of the free applications in Apple's iTunes Store could access users' contacts. In 2012, a class action lawsuit was launched against more than a dozen companies for selling mobile apps that uploaded users' contact lists without their knowledge or consent. Facebook announced in December 2011 that it would post archived user information, making old posts available under new downgraded privacy settings as part of a new Timeline feature. Users were given just one week to clean up their histories before Timeline went live. The extraordinary (and brazen) announcement came only a few short weeks after a decision by the U.S. Federal Trade Commission found that Facebook had engaged in "unfair and deceptive" trade practices when it changed the privacy settings of its users without properly notifying them.

Google's 2010 collection of private wifi data (described in the last chapter) was but one of several concerns users have had about the company's ambitious data collection practices. If a user employs the full range of Google products – from Search to the Android mobile operating system to Gmail, Google Docs, Google Calendar, Google Hangout, and others (all of which are free) – Google's consolidated management of the precise detailed information about each of its user's movements, social relations, habits, and even private thoughts is truly frightening in scope and scale, especially in the event that any of these capabilities is abused, compromised in some way, or subject to external controls and manipulation. Such a scenario is not far-fetched. In the 2009 Operation Aurora attacks Google's networks – including many Gmail accounts and some of the company's source code – were compromised by China-based attackers. After the attacks, Google entered into a secret agreement with the NSA to review its security. "The company pinkie-swears that its agreement with the NSA won't violate the company's privacy policies or compromise user data," wrote *Wired*'s Noah Shachtman, adding: "Those promises are a little hard to believe, given the NSA's track record of getting private enterprises to co-operate, and Google's willingness to take this first step." Critics were hardly mollified when the U.S. Electronic Privacy Information Center's (EPIC) freedom of information request to find out more about the secret agreement was rejected in May 2012 by a U.S. federal appeals court, which said that the NSA need neither "confirm nor deny" the existence of any relationship with Google. The world's largest data collection company secretly partnered with the world's most powerful spy agency, and no one outside of either institution knows the full details? It would be hard to conjure up a more frightening scenario.

Along with other social networking companies, Google has strongly resisted proposed European Commission regulations, colloquially known as the "Right to Be Forgotten," which would require companies to provide users with the option to have removed all user data they collect, including metadata. The "Right to Be Forgotten" legislation may never pass, but it does bring up a major set of issues surrounding the retention of data. Network operators and service providers vary in how long they retain the data they collect. Among mobile providers in the U.S., for instance, Verizon keeps a list of everyone you have communicated with through text messaging for twelve months, AT&T up to eighty-four months, Sprint for twenty-four, and T-Mobile for four to six months. The cellphone data that details a phone's movement history through its connections to cell towers and wifi hotspots is retained by Verizon and T-Mobile for twelve months, Sprint up to twenty-four months, and AT&T for an indefinite period of time. With apps, the data storage times are even more uncertain. When Twitter app users choose to "find friends," the company can store their address books for up to eighteen months. Most other apps say nothing at all about how they store user data, or how long they retain it.

Increasingly, laws regulate how long companies should retain data. The 2006 European Union Data Retention Directive makes it mandatory for telephone companies and ISPs to store telecommunications traffic and location data for law enforcement purposes for six to twenty-four months. All but three member states – Germany, the Czech Republic, and Romania – signed on to the law. One of its most egregious applications occurred in Poland. In its 2009 interpretation of the Directive, the Polish government gave its law enforcement and intelligence agencies the right to access data from private companies without any independent oversight, and without the government having to pay those companies compensation for the resources required to service those requests. Polish NGO Panoptykon found that Polish authorities requested users' traffic data half a million times more in 2011 than in 2010.

As the controversial EU Data Retention Directive suggests, the issue of what is done with all of the data we produce has become a critical public policy consideration. In a very real sense we no longer move about our lives as self-contained beings, but as nodes of information production in a dense network of digital relations involving other nodes of information production. All of the data about us as individuals in social network communities is owned, operated, managed, and manipulated by third parties beyond our control, and those third parties are, typically, private companies. In assessing the full spectrum of major social changes related to the information revolution, the entrusting of this unimaginably huge mass of civilian data in private sector hands ranks as perhaps the most important. As John Villasenor, a computer engineer at UCLA, puts it: "Most of us do not remember what we read online or wrote on April 29, 2011, or what clothes we wore that day. We don't remember the phone calls we made or how long we talked, or whether we went to the grocery store, and if so, what we purchased there. But all of that information is archived, and if a pressing enough need were to arise, our activities on that day could be reconstructed in nearly complete detail by third parties." All with our "consent."

Today, just because something happened in the distant past does not mean that it is forgotten, and there is no statute of limitations on our digital lives. It is all there, somewhere, possibly even copied numerous times over in multiple places – an endlessly proliferating series of duplicates of everything we do located in the deep recesses of servers, forever open to manipulation. Big Data meets George Orwell's Big Brother.

• • •

While the consumer big-data market touches us all directly, there is another world of big-data exploitation hidden in the shadows of military and intelligence agencies, and the many "fusion" and "analytics" companies that revolve around them. This other world of big data has its roots in 9/11 and in the perceived failure to "connect the dots" that led to that catastrophe: the widespread lament among U.S. defence, law enforcement, and intelligence personnel that the assault could have been prevented had the right people had access to all of the pertinent information. If only someone had been able to piece together emails and phone calls, a car rental, someone passing through a border checkpoint on a temporary visa from a country known to harbour terrorists, civilian flight school enrolment . . . .

The goal of preventing the next 9/11, and of rooting out shadowy terrorist cells in Iraq, Yemen, Afghanistan, and elsewhere, led to another information revolution: big-data analytics. The challenge was clear: to find a way to integrate seamlessly all of the disparate data sets out there but trapped in silos across government agencies and proprietary databases. One of the first attempts to address the "data fusion" problem ended up backfiring, at least in the public eye. The idea behind the Total Information Awareness (TIA) system, spearheaded by ex-admiral and CIA officer John

Poindexter, was simple: find a way to integrate as much data as possible about everyone – not only data that is classified, but also open-source information – into a single, searchable platform. Credit card transactions, tax records, flights, train trips, and numerous other pieces of information were thought fair game by Poindexter. His involvement in the 1980s Iran-Contra affair, coupled with the frightening civil liberties implications of what TIA embodied, provoked howls of protest among privacy advocates and this effectively short-circuited the project from getting the necessary Congressional approvals. But the TIA did not get shelved. According to Shane Harris, author of *The Watchers*, the TIA went "black budget" and was supported under a secret umbrella, one that was not subject to public scrutiny. TIA was driven underground, but most experts believe that it remained operative.

At the same time as the TIA went dark other projects were being seeded with the same goal, and the result today is a major new defence industry around data fusion and analytics. One of the chief driving forces behind this push was the CIA's investment arm, In-Q-Tel, which financed a number of startups in the data fusion and analytics arena. One of them, Palantir – a company whose origins lie in the PayPal fraud detection unit and whose founders were given early advice by Poindexter – has become a darling of the defence and intelligence community, but a bit of an outcast among civil libertarians. In February 2011, an Anonymous operation breached the networks of the security company HBGary, and then publicly disclosed plans they had uncovered involving HBGary, the Bank of America, Palantir, and others to attack WikiLeaks servers, and spread misinformation about Wikileaks supporters, including the journalist Glen Greenwald. Although Palantir's CEO apologized and then distanced his company from the misguided plan, the taint of the association still lingers among many. (Full disclosure: In 2008, Palantir donated a version of their analytical platform to the Citizen Lab, and we employed it only as a minor research tool during the GhostNet and Shadows investigations.)

Although an industry darling, Palantir is only one among a growing complex of data analysis companies that orbit the law enforcement, military, and intelligence communities. A two-year investigation by the *Washington Post*, called "Top Secret America," provides some startling insights: "Some 1,271 government organizations and 1,931 private companies work on programs related to counterterrorism, homeland security and intelligence in about 10,000 locations across the United States," and companies that work specifically on data analysis and in information technology include Alion Science and Technology, Altera Corp., BMC Software, Cubic Corporation, Dynetics, Inc., ESRI, Informatica, Mantech International, MacDonald, Dettwiler and Associates, Inc., Verint Systems Inc., and many, many others. It's a multi-billion-dollar annual business. These companies' systems are used to parse enormous databases, scour all existing social networking platforms, integrate data from the vast troves in the hands of telecommunications companies and ISPs, and piece it all together to provide decision makers with actionable intelligence. As former CIA director David Petraeus explained at In-Q-Tel's CEO Summit in March 2012, "New cloud computing technologies developed by In-Q-Tel partner companies are driving analytic transformation in the way organizations store, access, and process massive amounts of disparate data via massively parallel and distributed IT systems . . . among the analytic projects underway with In-Q-Tel startups is one that enables the collection and analysis of worldwide social media feeds, along with projects that use either cloud computing or other methods to explore and analyze big data. These are very welcome additions to the initiatives we have underway to enable us to be the strongest swimmers in the ocean of big data."

One of the more lucrative of these markets, and potentially the most troubling for privacy, is for biometrics and facial recognition systems. While developed for military, law enforcement, and intelligence purposes – approximately 70 percent of current spending – the broader consumer market is growing fast. Many social media and mobile platforms use facial recognition technology on their digital photo apps so that users can tag, categorize, and verify their own and their friends' identities – apps like Photo Tag Suggest, Tag My Face, FaceLook, Age Meter, Visidon AppLock, and Klik (produced by Face.com, an Israeli-based company that was acquired by Facebook for $100 million in July 2012). In 2010, Cogent, a company which provides finger, palm, face, and iris ID systems to governments and private firms, estimated that the biometric market stood at $4 billion worldwide. Cogent then expected this figure to grow by about 20 percent a year, driven mostly by governments and law enforcement agencies interested in identification systems, a forecast that appears to have been proven correct.

In 2011, Google Executive Chairman (and former CEO) Eric Schmidt explained that he was "very concerned personally about the union of mobile tracking and face recognition." Just the same, Google had purchased several facial recognition start-ups, including Viewdle and PittPatt, and the German biometric company, Neven Vision. In the same year that Schmidt expressed his concerns, Google launched an "opt-in" photo-tagging feature, Find My Face, for its social network. Just as the NASA space program partially justified its existence on the basis of civilian benefits and spinoffs (and we ended up with freeze-dried food and memory foam as a result), so too we can expect the military and intelligence big-data fusion market to find its way into the civilian marketplace. Regulators are beginning to take note. In 2012, the Hamburg Commissioner for Data Protection and Freedom of Information in Germany issued an administrative order to Facebook to cease its automatic facial recognition system until it could bring its operations in line with European data privacy policies. The question remains: Will regulations like this be enough to stem the tide?

• • •

It is easy to demonize the companies involved, but big data is related to our own big habits and big desires. It is we users, after all, who share and network through social media, and it is we who have entrusted our information to "clouds" and social networking services operated by thousands of companies of all shapes, sizes, and geographic locations. We are the ones who have socialized ourselves to share through clicking, through attachments and hyperlinks. Today, surveillance systems penetrate every aspect of life, and individuals can be mapped in space and time with an extraordinary degree of precision. All of this has emerged with our generally unwitting consent, but also with our desire for fame, consumption, and convenience. The *who* that makes up cyberspace is as important as the *what* – and the *who* is changing fast.

4.
The China Syndrome

On November 16, 2009, under the headline "UN Slated [sic] for Stifling Net Debate" and the subhead "The UN has been criticised for stifling debate about net censorship after it disrupted a meeting of free-speech advocates in Egypt," BBC News reported the following: "UN security demanded the removal of a poster promoting a book by the OpenNet Initiative (ONI) during a session at the Internet Governance Forum in Egypt. The poster mentioned internet censorship and China's Great Firewall. The UN said that it had received complaints about the poster and that it had not been 'pre-approved.' 'If we are not allowed to discuss topics such as Internet censorship, surveillance, and privacy at a forum on Internet governance, then what is the point of the IGF?' Ron Deibert, co-founder of the OpenNet Initiative, told BBC News."

It came out of nowhere. Our plan to hold a book reception for *Access Controlled*, the second in a three-part series of books on cyberspace, probably would have been yet another sleepy affair had Chinese authorities not intervened. Instead their intervention had a "Streisand Effect," a phenomenon, according to Wikipedia, "whereby an attempt to hide or remove a piece of information has the unintended consequence of publicizing the information more widely." Had the Chinese government not been censoring access to Wikipedia, they might have heard of the "Streisand Effect" and left the book launch alone. Alas, no.

In 2009, the annual Internet Governance Forum (IGF) meeting was held in Sharm el-Sheikh, Egypt. An incongruous location – the massive conference centre sits in the middle of the desert like a postmodern pyramid – but swarms of attendees from around the world descended on the facility, lanyards draped around necks, shoulders drooping from the weight of overstuffed conference tote bags. The IGF was set up by the UN as a forum to encourage multi-stakeholder discussions on Internet governance, discussions that included civil society groups and the private sector in an arena typically reserved for governments. It might as well have had a red target painted on it for those intent on preventing that very thing from happening.

Our book launch was disrupted before it even began by UN officials and security guards. First, they approached me demanding that we cease distributing pamphlets advertising the event, circling in pen a reference to Tibet in our promotional material. I was confused. "What's going on here?" I asked. After some back and forth, I agreed to stop distributing promotional material as the event was about to begin and the benefits of additional PR seemed negligible. Moments later, however, the security officials returned, this time insisting that we remove our banner displayed outside the room. An enlarged version of the book cover, the banner included back-of-jacket promotional blurbs and descriptions. When I asked what the problem was, an official pointed to the description of the "Great Firewall of China" and spoke in hushed tones about a "member state" issuing a formal complaint. "You can't be serious," I said.

The banner was brought into the room and inelegantly laid on the floor, the debate with security guards continuing as curious onlookers gathered around. Activists, diplomats, and scholars circled us, the number of people in the room growing as word spread. Inevitably, smartphones were pulled out, pictures and videos taken, most of them immediately posted on the Internet. (Many are still available online.) Though we protested vehemently, the banner was gathered up and escorted out of the room, followed by a procession of catcalls and mock applause. What I expected to be a non-event had turned into a political melee. Press inquiries began almost as soon as the book launch was over, and I found myself fielding calls from the BBC, CBC, and news outlets from around the world. "Did the United Nations censor your book launch?" "Did they really tear up your poster?" I was repeatedly asked. The IGF's president, Markus Kummer, made matters worse by excusing the shutdown as a problem not with what was printed on the banner, but with the banner itself. "Commercial banners are not permitted at the IGF," he claimed, his remarks plainly contradicted by dozens of other banners and posters spread throughout the facility. Kummer appeared less concerned about offending free speech than he was the Chinese government. Included in the crowd was the UN Special Rapporteur for Human Rights, Frank La Rue. A few months later, he issued a statement proclaiming the Internet "a human right." I wondered to what extent the events in that room inspired his actions.

Internet pundits like to think of autocratic countries like China as aging dinosaurs flailing about trying to stay afloat amidst the tsunami of digital information. In truth, while the West invented the Internet, gave it its original design and vision, and, latterly, assigned to it all sorts of powers and capabilities for rights and freedoms, what cyberspace will look like in the future and how it will be used will largely depend on *who* uses it. The Internet may have been born in Silicon Valley or Cambridge, Massachusetts, but its destiny lies in Shanghai, Delhi, and the streets of Rio de Janeiro, the places where its next billion users are going to come from. China's actions at the IGF may have been ham-fisted, but they were not an accidental or ad hoc reaction. Rather, they were part of a concerted effort to control cyberspace, an effort we all need to understand in its details.

●●●

"In America, the Internet was originally designed to be free of choke points, so that each packet of information could be routed quickly around any temporary obstruction," James Fallows wrote in the *Atlantic Monthly* in March 2008. "In China, the Internet came with choke points built in." Fallows is referring to the Great Firewall of China (GFW), the country's Internet censorship system, subject of the IGF banner controversy, and a descriptor that is both metaphor and not.

Secrecy surrounds the GFW but it is China's Internet backbone and guardian, the country's deepest layer of communications infrastructure through which all Internet traffic must eventually pass, specifically at three international gateways that connect China's Internet to the wider world: the Beijing-Tianjin-Qingdao connection point, in Shanghai, and in Guangzhou. For a country with more than 500 million Internet users surfing, texting, downloading, emailing, this is a small set of funnels, but the routers automatically inspect all traffic moving in and out, acting as a kind of border patrol. Requests for content that contains banned keywords, domains, or IP addresses are punted back unceremoniously. Unlike other countries that impose national Internet censorship regimes and that present back to the user a "blocked" or "forbidden" page, the Chinese system sends a wrench into the user's machine, a "reset" packet that disables the connection and sends back a standard error message giving the impression that the content requested doesn't exist ("file not found") or that something is wrong with the Internet. It's an ingenious way to frustrate users: if you make some websites persistently inaccessible, slow, or maddeningly unreliable for long enough, most people will eventually look elsewhere. Meanwhile, certain Chinese-based content is made widely and freely available for those who want to surf *a* Net, if not *the* Net.

What other functionalities are contained in these gateway routers – surveillance through deep packet inspection of email, for instance – is anyone's guess, but most cyberspace analysts suspect that the gateways are designed not just to block content but also to siphon up and monitor communications.

The GFW is part of an elaborate regime of domestic cyberspace controls, one element in China's overall information and communications strategy. It is reinforced by a thicket of laws, policies, regulations, and policing up and down the ecosystem of Internet communications. Contrary to principles of network neutrality, ISPs, hosting companies, websites, chat clients, and blogs operating in China are all required to police their networks. Internet cafés are routinely surveilled, and all individuals and organizations are held accountable by law for what they do and post online. According to a 2010 White Paper published by the Chinese government:

No organization or individual may produce, duplicate, announce or disseminate information having the following contents: being against the cardinal principles set forth in the Constitution; endangering state security, divulging state secrets, subverting state power and jeopardizing national unification; damaging state honor and interests; instigating ethnic hatred or discrimination and jeopardizing ethnic unity; jeopardizing state religious policy, propagating heretical or superstitious ideas; spreading rumors, disrupting social order and stability; disseminating obscenity, pornography, gambling, violence, brutality and terror or abetting crime; humiliating or slandering others, trespassing on the lawful rights and interests of others; and other contents forbidden by laws and administrative regulations. These regulations are the legal basis for the protection of Internet information security within the territory of the People's Republic of China. All Chinese citizens, foreign citizens, legal persons and other organizations within the territory of China must obey these provisions.

(If the Puritans suffered from a profound fear that someone, somewhere was having a good time, given these "provisions" what can we say about the Chinese government?)

China routinely downloads responsibilities to police the Internet to the private sector, which must follow government regulations in order to be allowed to operate. In 2008, the Citizen Lab discovered that the Chinese version of Skype, TOM-Skype, was coded in such a way that it secretly intercepted private (and encrypted) chats whenever people used any number of banned keywords – Tiananmen and democracy, to name two. Despite the outrage after the release of our report and the condemnation levelled at Skype for colluding with Chinese authorities, four years later the same system is still in place. In fact, it is now more elaborately designed and frequently updated, sometimes on a daily basis in response to current events like the ongoing dispute with Japan over islands in the South China Sea, or the controversy around disgraced Communist Party official Bo Xilai. In fact, all Internet companies operating in China – Baidu, Sina, Tencent QQ, Youku, and others – are required to stop the "spread of harmful information" over their networks. The policing is typically undertaken through filtering and surveillance of the type TOM-Skype engages in, enforcing the use of real names in registration processes (to eliminate anonymous postings), and even direct intervention by paid officials in forums warning users not to engage in unwelcome, perhaps even illegal, discourse.

While downloading control to manufacturers of equipment and services is routine in China, occasionally there is pushback. For example, a proposal to have all new PCs manufactured in China come pre-equipped with the Green Dam censorship system met with widespread condemnation from users and was withdrawn. However, though the Green Dam was a big "ask" even for the Chinese government, more often than not companies simply comply in order to do business.

The system is hardly foolproof. Researchers at Cambridge University, for instance, once demonstrated how easy it would be to disable the GFW, and even without outside meddling the gateway routers can be overwhelmed by peak usage. Also, technical means to circumvent the GFW are plentiful. Using tools like Tor, Psiphon (a circumvention tool invented in the Citizen Lab in 2006, and now a private Canadian company), and commercial virtual private networks (VPNs), many users play a cat-and-mouse game with authorities; by some estimates millions break through censorship walls on a daily basis. Chinese citizens have also proven themselves adept at outflanking and mocking the censors. Code words, metaphors, neologisms, and ingenious images circulated as Internet memes are used in place of conventional terms to circumvent Skype and other companies' filtering and surveillance regimes. So, when any reference to "Bo Xilai" was censored, Internet users began referring to him as "Gua's Father" instead (indicating that Bo Xilai is the father of Bo Guagua), until that term was filtered, and so on. The average Chinese user might go days without bumping into attempts of state control online, but the threat is always lurking. In this sense the system is less like *1984* and more like Jeremy Bentham's Panopticon, a system that gives the *feeling* of being watched, that someone somewhere knows what you're doing. No doubt, this creates considerable self-censorship, especially when combined with high-profile arrests of those who openly challenge the system.

It's noteworthy that China's cyberspace strategy – unlike, say, North Korea's – is not aimed at completely isolating the country's population from outside influence. Rather, it's deliberately designed to take advantage of information and communications technologies which the Chinese see as critical to their long-term future, while maintaining political stability around one-party rule. Continued economic prosperity is essential to the legitimacy of the Chinese Communist Party, and information and communications technologies are central to a burgeoning knowledge economy. China doesn't fear the Internet; rather it embraces its own particular version of it. Indeed, the Chinese are building a robust alternative design that may actually be succeeding.

• • •

Often ignored is the connection between China's domestic controls and the international dimensions of its cyberspace strategy. Part of China's objective is the wholesale exploitation of cyberspace for intellectual property, political espionage, and targeted threats against meddlesome human rights, ethnic, and religious groups the government describes as separatists or terrorists. It has pioneered ways to vacuum up information of strategic value to the government and national industries, and has done so without shame. The GhostNet probe may have been one of the first to expose what this looks like from the inside out, but it was neither singular nor unique. Evidence of GhostNet-like compromises now surface almost weekly, and show a level of audacity and rapaciousness that is remarkable: dozens of government ministries and departments, from spy agencies to prime ministers' offices in numerous countries, have been breached, with all the perpetrators having operated out of Chinese Internet networks. Scores of defence, aerospace, petrochemical, nuclear, and communications companies have all been infiltrated, and dozens of NGOs have had their computers turned into the digital equivalent of wiretaps.

A particularly noteworthy case is Canada's Nortel Networks, which was thoroughly compromised for nearly ten years. In 2012, ex-Nortel employee Brian Shields, who had led the forensic investigation of the compromise, came forward to disclose his experiences. According to Shields, the breach (which Shields traced back to IP addresses in China) was so thorough that the attackers had control of seven passwords from top company executives, including the CEO, which gave them complete and direct access to the company's internal secrets and intellectual property. (Attackers downloaded technical papers, R&D reports, business plans, employee emails, and other documents from computers under their control.) Shields discovered the breach in 2004, but his warnings were constantly ignored by top executives, one of whom (former CEO Mike Zafirovski) subsequently admitted that they just "did not believe it was a real issue." Shields estimates that the attacks had been going on since at least 2000, and lasted nine years. Nortel went bankrupt in 2009, and Shields's revelations have caused many to wonder about the possible connections between the breaches, its demise, and the rising fortunes of Nortel's chief China-based competitors, Huawei and ZTE.

In 2012, China's state-owned company, Sinopec Corp., made a controversial bid to acquire Talisman Energy, one of Canada's top oil and gas exploration companies, for more than $1.5 billion. While Canadian news reports focused on the question of foreign ownership of national assets, few noticed that Talisman Energy had been victimized by a major China-based cyber espionage operation called Byzantine Hades in 2011. The attackers gained access to Talisman's Asian-based networks, and had control of them for over six months. (Notably, a Bloomberg News report on this issue disclosed that the same Chinese attackers, called The Comment Group, had infiltrated the computer of a Canadian Immigration and Refugee Board adjudicator involved in the case of Lai Changxing, a Chinese tycoon extradited by Canada to China, where he is now serving a life sentence in prison.) There is no evidence connecting the hackers to the

Talisman takeover bid, but it certainly raises some intriguing questions about whether, and/or to what extent, information gleaned by the attackers made its way to Sinopec.

In 2001, three individuals working for the state-owned Datang Telecom Technology Company of Beijing were indicted for stealing secrets from U.S.-based Lucent Technologies. In 2002, two people funded by the City of Hangzhou were indicted for stealing secrets from several Silicon Valley technology companies, including Sun Microsystems and NEC Electronics. In 2003, an employee of PetroChina working with U.S.-based 3D-GEO was found to have copied up to $1 million of 3D-GEO's proprietary source code for seismic imaging onto his laptop. In 2009, an employee at Ford Motor Company was arrested and found guilty of stealing trade secrets on behalf of Beijing Auto. When such cases are combined with the reports of widespread China-based cyber espionage, it raises the question: Is it all part of a deliberate campaign?

While cyber theft and spying are menaces, the potential military implications are more frightening. It's unlikely that China would see any benefit in an armed conflict with the United States, but Chinese military literature emphasizes its capacity to degrade American satellites, as well as its other surveillance systems, should an armed conflict occur. Like those of many other countries, China's military planners have fully integrated cyber warfare into their military doctrine and operational plans. Because the U.S. has a military alliance with Taiwan and Japan, in the event of a regional war – say, over Taiwan or the disputed islands of the South China Sea – the People's Liberation Army would be hard pressed *not* to deploy its cyber warfare assets to confuse, deter, and even disable American military and civilian assets. As the Stuxnet worm aptly demonstrated in 2010, a menacing virus or trojan horse can be used to sabotage critical infrastructure. Such an attack would invariably provoke a wider response from the U.S., which now defines a cyber attack as an act of war. As security strategist Herman Kahn noted about the Cold War, this can be described as an "escalation ladder," one step leading to another, further and further into an armed imbroglio that neither side fully controls or desires.

Part of China's international strategy revolves around the setting of technical standards, like those relating to wifi protocols. In the early 2000s, after China lobbied unsuccessfully to have its WAPI standard for wireless networking adopted internationally, its government turned to promoting WAPI (WLAN Authentication and Privacy Infrastructure) as the domestic standard instead, making many handsets less than fully functional. For example, the official Chinese iPhone offered by China Unicom didn't include wifi (which helps explains the burgeoning iPhone grey market in the country). However, in 2010 Apple introduced a new-generation iPhone with the China-preferred WAPI wireless standard on its handsets, as did Motorola and Dell. In discussing such standards, it is noteworthy that Huawei is now the world's largest telecom equipment manufacturer, bypassing Sweden's Ericsson in 2011, and China's Lenovo is now the second-largest PC maker in the world, behind only Hewlett-Packard. Technical standards are the *sine qua non* of cyberspace control: they shape the realm of the possible, structure the limits of what is permissible, and define a path of dependency for future trajectories of technical development that is difficult to escape. When millions of devices worldwide contain a particular country's standards, those devices are linked to that country's industry and manufacturing base, and contain a structure of rules that are set by the decisions of that one country.

• • •

While technical standards-setting may work in indirect ways to further China's influence abroad, its policy engagement at regional and international forums is more directly illustrative of its determination to control cyberspace. China's participation at international forums where global cyberspace rules are debated has grown significantly, its agendas more clearly articulated and promoted. The country's representatives come in force, usually well prepared and organized around a common agenda at the Internet Corporation for Assigned Names and Numbers (ICANN), the Internet Engineering Task Force (IETF), the International Telecommunications Union (ITU), the UN Group of Governmental Experts on Cyber Security, and, as we discovered, the Internet Governance Forum. Their actions at the book launch may have been Monty Pythonesque, but the long-term effect of China's influence on the IGF is anything but laughable. Presently the IGF reports to the UN Department of Economic and Social Affairs, whose top person just happens to be Mr. Wu Hongbo of China.

China is also active at a regional level, as evidenced by its leadership, along with Russia, in a security alliance called the Shanghai Cooperation Organization (SCO). The SCO also includes Kazakhstan, Kyrgyzstan, Tajikistan, and Uzbekistan. Afghanistan, India, Iran, Mongolia, and Pakistan have observer status, and Belarus, Sri Lanka, and Turkey are dialogue partners. The organization is used to coordinate security concerns, primarily through the Regional Anti-Terror Structure, known by its acronym RATS. At RATS and SCO meetings member states' security services coordinate anti-terror exercises and share information on "threats" – which many human rights groups suspect include domestic opposition groups. Transparently, the intent is to restrict citizen-led revolts like those of the Arab Spring, tied as they were to social media. At a 2012 meeting of the RATS, Sergei Smirnov, first deputy director of Russia's secret service agency, the FSB, said: "New technologies are used by Western secret services to create and maintain a level of continual tension in society with serious intentions extending even to regime change . . . Our elections, especially the presidential election and the situation in the preceding period, revealed the potential of the blogosphere."

The information and technology sharing that goes on through RATS and SCO demonstrates a clear trend: the global surveillance industry is reaping the benefits of regimes with intense cyber-security anxiety. American, Israeli, and Canadian companies, as well as their Russian and Chinese competitors (with close associations to ruling elites), are all inserting their surveillance products and services into the systems of control that SCO is helping to propagate across the region.

To think that the Internet embodies some kind of magical formula to resist the full weight of these pressures, let alone dissolve them upon contact, is ludicrously idealistic. The more sensible question to ask is: What will the Internet look like when the next billion users, most of whom are located in the global South and East, come online?

5.

The Next Billion Digital Natives

Somalia has not had a properly functioning government since 1991, when the country began tearing itself to shreds in a brutal civil war. More than twenty years later, warring factions continue to fight over territory, people, foreign aid, and revenues from illicit activities. This quagmire has made the country one of the world's most dangerous places, and yet Somalia also has one of the most efficient and affordable cellphone infrastructures in Africa, if not the world. In the midst of this hardscrabble, anarchic place, there is not one but four independent and thriving cellphone providers: Hormuud, Golis, Nationlink, and Telesom. Though competitors, they connect seamlessly and offer access to most areas of the country. Somalia has the lowest international call rates in Africa, a monthly fee of just $10 gets you unlimited local calls, and the wait time for a new land line is typically only a couple of days, compared to as much as three years or more in other African states. Launched in 2010, Somalia's mobile banking service, ZAAD, allows users to transfer funds, make purchases, pay bills, and share airtime credits with friends and family. Not surprisingly, the number of Somalis connected to the Internet has exploded from virtually nil twenty years ago to almost 40 percent of the population today.

How to explain this paradox? There are places in the world where the darker sides of human nature seem to roam free:

Mogadishu, Grozny, Ciudad Juarez, Karachi, to name a few. Here, life can seem tenuous and cheap, the hallmarks of civility and the social glue that Westerners take for granted illusive. Violence is pervasive, and gangsters fill the void left by the absence of legitimate public authority. In Chechnya, a nation of roughly 1.3 million, there are eighty-four murders a day. The UN ranks Somalia 181st out of 194 countries for life expectancy, a mere forty-eight years. But while the bursting cities of the developing world – flush with *favelas*, with streets that reek of diesel and are littered with garbage, with omnipresent poverty and crime just around the next corner – may seem inhospitable to the digital world, they are not. Although the digital divide still exists, it's shrinking, fast: cheap and easy-to-use means of communication are as irresistible here as anywhere on Earth, and as digital networks approach 100 percent saturation in the global North, the South and East have become the next frontiers of rapid expansion. Somewhat counterintuitively, the persistent chaos, corruption, and crime might facilitate, rather than retard, their wider development.

While these new digital natives love their smartphones and apps as much as the rest of us, the uses to which they are often put are dramatically different, and this difference is affecting cyberspace itself. If cyberspace in the 1990s was a result of the collective dreams of Silicon Valley upstarts, today it is increasingly the unfulfilled aspirations of teenagers growing up in the endless grey apartments of Moscow, the cramped cubicles of Guangzhou, the teeming slums of Rio de Janeiro that are its principal drivers. It is a Western conceit to think that just because we invented the technology its uses will conform to our original design. As Somalia and many other cases show, human ingenuity is never so predictable, nor is the path of social evolution so linear.

• • •

There are at least three reasons for Somalia's robust cellular infrastructure, all of them directly related to the disastrous civil war. When the country's central government collapsed, so did "rent seeking," a term coined by the economist Anne Krueger in 1974 to describe the actions of companies and/or governments to extract revenues by using their positions of market authority to raise prices ("rents"). State-owned telecommunications companies have been particularly prone to rent seeking: artificially inflated rates applied to long-distance calls have been used by governments the world over to extract revenues for the state. In developing countries, the situation was dire. For decades – with the exception of those countries that underwent privatization in the 1990s – laws were passed to keep out competition because the telecommunications sector, it was argued, constituted a "natural monopoly." In Somalia, when the central government collapsed, so did the underpinning for regulations, approvals, permits, and taxes, and into this void leapt private telecommunications entrepreneurs.

A second reason involves the competing armed factions and tribal groups. Although viciously competitive, the one thing that all of them could agree on was the need for a working telecommunications infrastructure. Militants needed a functioning Somali-wide cellphone system to operate their enterprises, issue threats through text messaging, make payments for drugs and weapons, and coordinate their tribal factions. Mobile phones were also less easy to intercept than land lines and provided a degree of anonymity, as the phones and their SIM cards could be recycled and exchanged. With disposable phones and SIM cards, there is not the same registration process that fixes one individual to a specific phone (in the way that there would be with a land line), making tracking of communications more of a challenge (though not impossible). Some Somalian cellphone companies are rumoured to be tied to militant factions, in particular one of the biggest providers, Hormuud, and the al-Qaeda–linked Somali jihadist group al-Shabaab. (The latter has a curious relationship to telecommunications: it has accused those who use cellphone scratch cards of being un-Islamic.)

Hormuud, it turns out, profits when attacks are on the rise in areas that al-Shabaab controls. When fighting erupts, people are afraid to go out and call friends and relatives at home or abroad to reassure them about their safety. In al-Shabaab–controlled areas, young men and women are forbidden from socializing, and tend to text each other instead. With danger lurking outside, people stay in and use their phones. The thriving mobile banking service is also partly a function of the perennial violence in Somalia: being able to make purchases, deposits, and transfers electronically is much safer than carrying hard currency in a place where theft and kidnapping are daily occurrences. In all, Somalia's functioning telecommunications system defies conventional political economic theory: a public good that works without a central authority and that provides a degree of security.

Finally, as the country was being devastated from the inside out, Somalis fled to other parts of the world in droves, and there is now a huge Somali diaspora population in Europe and North America. As a rule, Somalis abroad tend not to forget their friends and relatives, and Somali cab drivers, nurses, teachers, engineers, and so forth routinely send remittances to their home country through the informal *hawala* banking network. Estimated at $1 billion annually, remittances are critically important to the Somali economy, representing as much as 50 percent of GDP, making Somalia the fourth most remittance-dependent country in the world. Somalia has no central bank, nor many functioning commercial banks, and thus *hawala* firms are the only real vehicles of economic exchange and income for many households and businesses. Telecommunications is essential to *hawala* networks.

How does it work? A Somali in, say, Toronto visits a local *hawala* broker, a *hawaladar*, and gives him an amount of money destined for a relative in Mogadishu. The *hawaladar*, in turn, communicates this transaction to a Somali-based *hawaladar*, who brokers the deal and keeps track of the amount, usually by email or Skype. The transactions are entirely based on trust and the promise of repeat business. A small percentage fee is charged to the sender, who authorizes the transaction by communicating a password to the intended receiver. For years, the *hawala* network relied on telephone land lines and telegraphs, but today it is the Internet and devices used to connect to the Internet, like cellphones, that are central to *hawala* exchanges. Indeed, the final part of the transaction is almost always done online, and one of the primary uses of cellphones in Somalia is mobile banking.

• • •

In 2012, when the U.S. attempted to shutter *hawala* financing because of suspicions it was funnelling funds to al-Qaeda, the sanctions had a direct impact on Somali communications and inadvertently helped to establish one of the biggest cellphone providers, Hormuud. One of the owners of Hormuud, Ahmed Nur Ali Jim'ale, was also once a part-owner of the al-Barakat money transfer company targeted by U.S. and UN sanctions. Prior to setting up Hormuud, Jim'ale ran both al-Barakat and its subsidiary, Barakat Telecommunications, and he and both companies were placed on the Special Designated Terrorist Group list by the UN Security Council in 2001. In response, Jim'ale was part of a group that founded Hormuud Telecom in April 2002.

Ten long years later, Jim'ale and the entire al-Barakat conglomerate were finally removed from the list of al-Qaeda–related terrorist organizations. But a separate UN Security Council committee then put Jim'ale on a new list, one that imposed a travel ban and froze his assets, the suspicion this time being that he was an arms trafficker. In its decision the Security Council noted that "Jim'ale established ZAAD, a mobile-to-mobile money-transfer business and struck a deal with Al-Shabaab to make money transfers more anonymous by eliminating the need to show identification"; that his company, Hormuud, is "one of the single largest financiers of Al-Shabaab"; and that Hormuud "cut off telephone service during Al-Shabaab attacks against pro-Somali Government forces." Regardless of whether and to what extent Jim'ale and Hormuud have ties to al-Shabaab – or, in the past, had ties to al-Qaeda – the case shows the depth and complexity of Somalia's cellphone system. A country wracked by civil war for over two decades nonetheless has extraordinarily sophisticated and innovative online banking and cellphone services: so sophisticated that the security-obsessed are watching, watching closely.

• • •

The Somalia case defies conventional wisdom about technical innovation in failed states. Precisely because of Somalia's chaotic and violent character, along with its people's adaptive culture of trust and information sharing, an efficient cellphone infrastructure took root and flourished. It also demonstrates how the evolution of cyberspace is highly contingent on local circumstances. How and to what end a cellphone is used in Manhattan or Berlin can be entirely different from how it's used in Mogadishu or Lagos, or the countless other cities of the global South and East from which the majority of future digital natives will emerge.

We may not know exactly what the impact of the next billion users is going to be, but one thing is certain: they're coming online fast. According to a 2011 International Telecommunication Union

(ITU) report, developing countries increased their share of the world's Internet users from 44 percent in 2006 to 62 percent in 2011. Asia alone now has nearly half of the world's Internet-using population, with North America and Europe combined at less than one-quarter. The gap in penetration rates between the developed and developing world – how much of their respective populations are online – is also closing. New broadband penetration in the North is down to about 5 percent a year, the ITU report states, while in the South the annual growth rate is 18 percent. At the same time, only roughly one-quarter of the developing world was online by the end of 2011: that is, there is a still a huge percentage of the population yet to be connected there, and connect they will.

Some of the fastest growth rates in social media usage are in the Middle East and North Africa, no doubt inspired by the prominent role played by Facebook, Twitter, et cetera during the Arab Spring. According to the *Arab Social Media Report* series, which counts users of social media in the Arab World, the number of Facebook users nearly tripled in one year, from 16 million in June 2011 to 45 million by June 2012. Egypt alone constitutes one-quarter of all Facebook users in the region, and Arabic is the fastest-growing language used to communicate on Twitter. The 2011 *Arab Social Media Report* revealed that there were sixty-seven tweets a second in the Arab region.

The vast majority of Internet users in the global South and East are young. The market research firm Nielsen found that Chinese between the ages of fifteen and twenty-four are the heaviest mobile Internet users, with 73 percent of them reporting having used mobile Internet connections in the last thirty days, compared with only 48 percent of their U.S. counterparts, and 46 percent in the U.K. Youth comprise approximately 70 percent of Facebook users in the Arab region, and the UN estimates that the number of youth living in developing countries will grow by 90 percent by 2025. (Over 65 percent of the Middle Eastern and North African population is under thirty-five.) Young people are early adopters everywhere, but in the developing world they face much higher rates of unemployment – as high as 25 percent in the Middle East and North Africa. The combination of youth, unemployment, and radicalism is a well-documented concern, one that some analysts (such as the former managing director of the International Monetary Fund, Dominique Strauss-Kahn) have called a "time bomb" that will soon ignite. To this mix we must add another ingredient: instant connectivity to global computer networks.

Net-savvy, disenfranchised youth fuelled the protests that swept across the Middle East and North Africa during the Arab Spring, toppling dictatorships that most thought were immovable, but there is a more discomforting scenario, especially as the dust settles and the "youth bulge" looking for jobs grows restless. Just as Willie Sutton famously remarked when asked why he robbed banks – "because that's where the money is" – the temptation to breach servers that host valuable data thousands of kilometres distant but only a mouse click away might be irresistible to scores of computer-literate unemployed youth from Rio, Rangoon, or Marrakesh.

Perhaps the most noteworthy fact is that the fastest growth rates are occurring among the world's failed and most fragile states. In the ITU's 2009 Information Society Statistical Profiles, the ten countries that saw the fastest Internet user growth rates over the previous five years were Afghanistan, Myanmar, Vietnam, Albania, Uganda, Nigeria, Liberia, Sudan, Morocco, and D.R. Congo. (Afghanistan topped the list with growth in Internet users of 246.6 percent from 2002 to 2007.) The fastest growth rates in mobile cellular subscriptions over five years occurred in some of the world's poorest and most conflict-prone states: Guinea-Bissau, 230.2 percent; Afghanistan, 184.6 percent; Nepal, 172.2 percent; Ethiopia, 128.1 percent; Chad, 94.5 percent; Angola, 80.9 percent; and Burkina Faso, 60.7 percent. These statistics are almost certainly underestimations of the actual number of users connecting to cyberspace in these countries (gathering statistics in zones of conflict and in failing and fragile states being inherently difficult and almost always unreliable), but what we do know from field research in the global South and East is that users there improvise in ways that numbers do not capture: one or several families may share an Internet connection, or regularly use a shared Internet connection point at a café or corner store; cellphones may be shared between several people, multiplying the actual users represented by a single data point. Granted, many of these countries are starting out from a baseline of practically no connectivity whatsoever, but this does not change the fact that they are migrating online and into cyberspace at a furious pace, and in a context of chronic unemployment and minimal government control.

What to expect from these next billion users is hard to say, but innovative uses of cellphones, Internet cafés, satellite uplinks, and websites in the global South and East challenge our assumptions of the type of social effects spawned by cyberspace. The domain may be liberating, but liberating for what, exactly? Over the past ten years of research, much of which I have spent on journeys in poor, less developed regions of the world, I've often had a *Wizard of Oz* feeling of "not being in Kansas anymore."

• • •

In many parts of the developing world, the separation between organized crime and the state is blurred, meaningless really. Public officials use their offices for graft, or employ criminal groups to exercise paralegal authority. It should come as little surprise, then, that the levers of power in such places are used to control cyberspace.

The most striking examples come from the former Soviet Union, a ring of nations forcefully unified under Stalin's regime and sharing its legacy of controls: Russia and the "stans" (Kazakhstan, Kyrgyzstan, Tajikistan, Uzbekistan, Turkmenistan), Ukraine, Belarus, Georgia, Moldova, Azerbaijan, and Armenia. Whereas in other parts of the world cyberspace controls are exercised through technical means like filtering software, in these countries smashed windows, threatening phone calls in the middle of the night, broken bones, arrests, even murder are the not-so-subtle means of shaping the communications space. In Uzbekistan, the regime of President Islam Karimov has long used severe punishments to create a chilling effect on dissent: for instance, two political prisoners accused of belonging to an Islamic extremist group were executed by being boiled to death in 2002. Vladimir Putin's regime and others like it govern through a combination of intimidation and exploitation, making the boundaries between what is legal or criminal largely abstract, irrelevant in practice. Their political control strategies often veer into outright thuggery, and some organized crime groups might be better described as informal agencies of the state: political authorities regularly use their services to quell unrest or to further kleptocratic ends. Such stone-age techniques have been applied in the digital arena with nearly gleeful indiscretion by Eurasian countries.

Over the years, researchers in the region with whom we collaborate (often at great risk to them) have uncovered countless anomalies that point directly to the vested interests of entrenched authorities: the disabling of access to opposition websites leading up to critical elections; tampering and manipulation of DNS records to favour local, approved websites over international ones; political bloggers arrested on trumped-up charges of copyright violation or possession of child porn; brazen murders of critics of the regime. In one horrifying episode that caused me deep personal grief, a young Kyrgyz journalist named Alisher Saipov with whom Citizen Lab had collaborated was murdered in the middle of a busy street in

Osh, shot several times after what many believe was a $10,000 bounty put on his head by Uzbek security services (Saipov had written articles critical of Uzbek authorities). While reading the news of his death on Radio Free Europe's website, I noticed that his Skype account, over which he and I had communicated, turning green, signalling that he was online. I was even more taken aback when a chat message popped up: "Professor Deibert, how are you today?" I did not answer, knowing that it was not Alisher, who would never have addressed me in such formal terms. Citizen Lab researchers set up a honeypot computer, to lure whoever had Alisher's computer and Skype account into contacting us, perhaps giving away who was behind his murder. I opened his Skype chat window and sent a message with the honeypot link, asking the person on the other end to check it out. I never got a reply.

Sometimes the controls used in cyberspace in the former Soviet Union have little or nothing to do with politics; instead, they are just about personal financial gain. In the course of our research, we regularly came across locally hosted versions of websites masquerading as Google, wherein the domain name system would redirect requests for the legitimate Google site to one that was a reasonable facsimile in order to capture advertising revenues. In Uzbekistan, three of four ISPs uniformly blocked access to content, while the fourth was entirely filter free. Further field research uncovered that the owners of the fourth ISP were connected to the family of Uzbek president Islam Karimov, and were operating a filter-free service to capture revenues.

• • •

Each country in the global South and East deals with cyberspace challenges in unique ways. India, the world's largest democracy, is confronted by most of the usual challenges afflicting the developing world: sectarian and religious strife, overpopulation and unemployment, infrastructure decay, regional tensions – along with barely contained hostilities with neighbouring Pakistan over the disputed territory of Kashmir. Cyberspace policy issues vaulted to the top of India's national security agenda after the Mumbai terrorist attacks: three consecutive bombings in July 2008 that left twenty-six dead and hundreds injured. It was widely reported that those responsible coordinated their activities through disposable cellphones and forged SIM cards. Alongside revelations that India's national security establishment had been thoroughly breached by Chinese-based hackers and constant concerns over inflammatory Internet content offending various religious and cultural sensitivities, these attacks provoked a sudden urgency among Indian policy-makers to do something – anything – to control cyberspace and its now 100 million Indian Internet users. The result has proven extreme, draconian, and chaotic in its effects.

The country's Information Technology (Intermediaries Guidelines) Rules of 2011 place extraordinary policing responsibilities on ISPs and other services that operate in cyberspace. Companies are required to screen content and any Indian resident can compel Google to remove material he or she deems offensive. Content forbidden by the state includes anything that is "grossly harmful, harassing, blasphemous, defamatory, obscene, porno-graphic, paedophilic, libelous, invasive of another's privacy, hate-ful, or racially, ethnically objectionable, disparaging, relating or encouraging money laundering or gambling, or otherwise unlaw-ful in any manner whatever; or threatens the unity, integrity, defence, security or sovereignty of India, friendly relations with foreign states, or public order or causes incitement to the commission of any cognisable offence or prevents investigation of any offence or is insulting any other nation." One might wonder what *is* allowed in cyberspace in the world's largest democracy?

In December 2011, the Indian government asked Google, Microsoft, Yahoo!, and Facebook to set up a proactive "prescreening system" to look for objectionable content and remove it before it goes online. Meanwhile, Section 69 of the 2008 Information Technology Act gives the government the power – in the interest of the sovereignty, integrity, defence, or security of India – to direct any Internet service to block, intercept, monitor, or decrypt any information related to these areas. Failure to comply with such demands can lead to fines and up to seven years in jail for executives. Not surprisingly, companies have found it nearly impossible to meet such broad and unusual requirements, and both Facebook and Google are now facing criminal charges in India for not removing content. India has also waged a persistent campaign to require companies operating there to assist in surveillance, most notably Canada's Research in Motion (RIM), the maker of the popular BlackBerry mobile device. Several times India has threatened to expel RIM from the country if the company fails to comply with its demands. At the municipal level, Delhi city authorities have even launched an ambitious program to monitor every single individual visiting the city's cyber cafés.

• • •

In many countries of the global South and East the rule of law is unevenly applied and arbitrarily enforced, and the newness of the challenges presented by cyberspace cause governments to overstep or awkwardly apply regulations in the face of emergencies or crises. In response to violent demonstrations that erupted across his country against a U.S.-made anti-Islamic film and the publication of French cartoons of the Prophet Muhammad, the Pakistani minister of the interior, Rehman Malik, ordered all cellphone networks disabled, cutting off access to approximately 100 million users. The ban affected practically everyone, including private security guards, emergency responders, NGOs, and doctors who scrambled to find ways to communicate during the state-imposed blackout. Meanwhile, the Indian government banned all mass text messaging after false alarms about imminent attacks on minority groups circulated over SMS, leading to an estimated 15,000 people fleeing various cities in panic.

In Kenya, in an attempt to prevent the sale and distribution of cloned and pirated mobile phones, the government ordered ISPs to turn off nearly 2 million cellphones whose hard-coded numbers didn't match databases. Thousands of Kenyans woke up to find their phones suddenly didn't work, even many who insisted that their purchases were legitimate. In 2010, Turkey ordered ISPs to block access to YouTube, the regulation implemented in such a way that numerous other Google services were also blocked, including Google Books, Google Pages, Google Docs, and Google Translate. In 2005, testing by the OpenNet Initiative found that when South Korea put in place regulations to block several dozen pro-North Korean websites, they also impeded access to more than 3,000 completely unrelated websites that shared the same IP address as the pro–North Korean websites because they used the same hosting company.

More so than in the largely secular West, religion remains a motive force across the global South and East and a major influence on law and politics. This will invariably shape cyberspace as regimes in those regions transplant laws applied to traditional media to ISPs, cellphones, and social media, banning content that offends religious or cultural sensitivities and downloading policing to the private sector. In his 2010 report, *In the Name of God*, the Citizen Lab's Helmi Noman analyzes how "the flow of information in cyber-space in majority Muslim countries mirrors, to a large extent, the flow of information in 'real' space in these nations." Many majority

Muslim countries criminalize the promotion of non-Islamic faiths among their Muslim citizens offline, and they have now taken steps to ensure that the same laws are applied to cyberspace.

Islam is mentioned explicitly as the state religion in almost all Arab countries, and sharia law is a strong influence over the legal code. As a consequence, press and publication laws that make it a criminal offence to insult Islam are being carried into cyberspace. The terms of service of Oman's Omantel and Yemen's Y.net, for instance, specify that users refrain from using these services to contradict religious values. Saudi Arabia's Commission for the Promotion of Virtue and Prevention of Vice, a religious police unit in charge of enforcing sharia law, published a document on its website entitled "The Moral Vice of the Internet and How to Practise Hisbah" (*Hisbah*, roughly translated, means encouraging moral virtues while suppressing vice.) Likewise, access to homosexual content online is restricted in many Muslim countries because gay relationships are considered taboo, and are in many cases illegal.

While many of these restrictions are imposed by the state, Noman details how pressures to enact and enforce such laws often come not from governments but from civil society groups and/or religious leaders. In July 2012, Ra'if Badawi, editor of the Free Saudi Liberals website, was arrested under Saudi Arabia's Anti-Cyber Crime Law for violating state values by providing an online platform for people to debate religion. The primary impetus for the charges seemed to emanate from a powerful Saudi cleric, Shaikh Abdul-Rahman al-Barrak, who declared Badawi an "unbeliever . . . and apostate who must be tried and sentenced according to what his words require."

Although the protection of religion is generally put forward to justify such charges, the real purpose, many argue, is to silence political dissent. On point, our research suggests that similar conflicts are cropping up with increasing frequency across the Muslim world, particularly in countries facing social unrest. Recently, a Saudi religious cleric objected to women using emoticons, insisting that "a woman should not use these images when speaking to a man who is not her 'mahram' [husband, roughly translated] because these faces are used to express how she is feeling, so it is as if she is smiling, laughing, acting shy and so on, and a woman should not do that with a non-mahram man." These religious pressures have pushed many Islamic countries towards attempting to build a *halal* (permissible under Islamic law) Internet that would cordon off their populations through both technical and regulatory means. In 2012, for example, Iran took several steps towards the creation of its own *halal* Internet, creating new laws, censoring foreign websites, outlawing circumvention tools, and even, for a time, blocking access to all Google services.

In the Far East, the same pattern is emerging. In Thailand, insulting the royal family is forbidden, an ancient crime known as *lèse-majesté*. The law is enforced by the Office of Prevention and Suppression of Information Technology Crimes, which systematically scans the Internet looking for evidence of violations. Thailand threatened to censor all of YouTube, before it reached an agreement with the service to block only those videos that offend Thai law originating from its own jurisdiction. Nonetheless, critics claim that tens of thousands of websites are routinely censored in Thailand, and the law is widely seen to have been applied selectively, often in very harsh ways. In a notorious 2011 case, sixty-year-old Ampon Tangnoppakul was sentenced to twenty years in prison for sending four text messages deemed insulting to the queen.

In an echo of elsewhere, the Thai government downloads policing responsibilities to ISPs and website administrators. Chiranuch Premchaiporn, the former webmaster of a popular Thai website, Prachatai, is on trial and faces a prison sentence of up to fifteen years, not for anything she said or posted, but for not deleting comments fast enough from the web forum that she operates. In such a climate, website operators are going to err on the side of caution, creating a wide chilling effect. Those who question the journalistic ethics of online anonymity ought to look more closely at such cases.

• • •

Netizens in the West are used to thinking of either the state or corporations as the biggest threat to rights and freedoms online, but in many developing countries non-state and non-corporate actors pose the greatest danger. Although cellphone use is spreading in Afghanistan at a lightning pace, it operates in a regulatory space largely shaped by the Taliban. In the provinces and cities where they dominate, the Taliban has issued threats to cellphone providers to turn off signals at the towers they control, making cellphones unusable. Taliban leaders view this as a defensive measure to prevent informants from calling in Taliban locations to American forces or to keep NATO signals intelligence from eavesdropping on their communications. "Our main goal is to degrade the enemy's capability in tracking down our mujahedeen," a Taliban spokesman told the *New York Times*, but as the *Times* report noted, the motivation is just as likely to remind the population that the Taliban, not the government, has control over communications.

In Latin America some of the most innovative uses of information and communication technologies, and the most repressive in terms of their societal effects, have been deployed by drug cartels. In several high-profile cases, authorities have seized cartel-related assets that demonstrate advanced digital and networking capabilities – a giant 100-metre transmission tower, for instance! Today's cartel member is as heavily wired as a Palo Alto undergraduate student or Manhattan bond trader. Popular YouTube videos in Mexico glorify the lifestyles of the drug trade and are used to issue threats against rival gangs, and intimidate police and the general public. One of the most alarming innovations has been the use of the Comments section of YouTube video postings to communicate threats from one cartel to another, or from a cartel to the general public. The cartels have also shown a ruthless ability to employ social media to intimidate watchdogs and others from using those very same tools. One blogger critical of the cartels was beheaded, and then, in a macabre display, the cartels posted a video with her head on a keyboard, next to a cardboard sign warning others not to do the same. Cartel members who murdered a moderator of a social network left a note next to his corpse saying, "This happened to me for not understanding that I shouldn't report on the social networks." In another case, a sign attached to two dead bodies hung from a pedestrian overpass read, "This will happen to all the Internet snitches." In yet another case, disembowelled and mutilated corpses were hung from a bridge with a sign that read, "This is going to happen to all those posting funny things on the internet, you better fucking pay attention. I'm about to get you." Gruesome videos of informants or captured police officers being executed are set to music glorifying one or another gang, some of whom have become popular on local radio, probably as the result of intimidating broadcasters, or of outright ownership of media outlets by the cartels.

Most of us are vaguely aware that there is a seamier, darker side to the Internet, but we tend to assume it is hidden deep in the shadows. While that's true in some cases, in Mexico (and spreading through other parts of Latin America) the gruesome violence of the cartels is on full display, thriving in the new social media environment, while simultaneously presenting an extraordinary threat to freedom of expression. Barely noticed by the technorati of the industrialized North, Mexico is undergoing its own social media revolution, and it is having a regional, if not global, impact.

In February 2012, Google broke ground on an experimental super-fast fiber-optic network it launched in Kansas City, Kansas. The city was selected from a list of more than a thousand other American municipalities as the test ground for a one-gigabit-per-second Internet connection that would offer speeds up to 100 times faster for downloads and 1,000 times faster for uploads than a typical U.S. connection point. The Google manager behind the project, Kevin Lo, said that engineers had been busy planning, surveying, and eating "way too much barbecue."

At roughly the same time, thousands of miles away, two separate freak accidents resulted in the severing of four submarine cables to the African continent, shutting off connectivity to at least nine countries. Ten years ago, there was virtually no Internet access in Africa outside of South Africa and parts of North Africa, and there would have been, as a consequence, no cables to sever, no outages rippling across the region. Since then, the situation has changed, and dramatically so. Although there remain huge regulatory, energy, computer literacy, and other roadblocks, the continent is rapidly coming online, with growth rates approaching 2,000 percent per year, compared to roughly 480 percent for the rest of the globe. Africans, like other populations of the global South, have a growing stake in cyberspace.

Real communities rarely, if ever, emerge by fiat, or by any other artificial means. Rather, they coalesce organically, a result of individuals and interests growing together. Cyberspace may well be a global technological artifact, but it is colonized and inhabited by individuals and communities who have come together spontaneously, empowered by digital technologies and collectively creating its social ethos.

In the 1990s, the users and creators of cyberspace were largely white, prosperous, and clustered in the industrialized North and

West. By the mid-2000s, this was no longer the case. While the highest penetration of users remained in North America and Europe, the bulk of Internet users had shifted south and east. By 2012, two-thirds of all Internet users were located outside of North America and Europe, and over one-quarter were in China.

Images and metaphors of cyberspace are a useful way to portray its dominant characteristics. William Gibson, the science fiction writer who coined the term *cyberspace*, paints a picture of the domain as a virtual-reality matrix in which users physically plug their minds into a world of "endless city lights receding." The image evokes clean spheres and precise mathematical coordinates, like the contours of 3D computer graphics. Gibson was influenced by his experiences of the game arcades that lit up Granville Street in downtown Vancouver, where he lived. For many cyberspace users today this consumerist abstraction is still the dominant impression.

For the next phase of its evolution, however, the more appropriate image would perhaps be the *favela*, or shantytown, which better describes where the next billion cyberspace users will come from. Most of the next billion digital natives will be under twenty-five years old, and most will live in societies where the chances for local prosperity are relatively slim, and where the political institutions tend to fall on the authoritarian end of the spectrum, if there are political institutions at all. To them, the glittering virtual realities of cyberspace represent a world far removed from their own – a world of wealth, opportunity, boundless creativity, and hope. It is in these back streets of the developing world, with their crowded Internet cafés and burgeoning wireless access points, that the future of the Internet is now being forged.

Western states came late to the game of governing the Internet. Indeed, there were deliberate policy choices made early on to "keep the state out." Al Gore may not have "invented the Internet," but he did play an important role in Congress by defining a limited role for government in structuring how it should be regulated. Not so for most governments in the South and East. They approach cyberspace at a time of heightened concerns around cyber security, where threats are everywhere, and states like Russia and China are offering solutions.

We tend to think of globalization as a torrent of ingenuity originating in the industrialized North and West and spreading outwards. But globalization is a two-way process. The same networks that spread information from London, Tokyo, and New York offer a channel in the other direction from the developing world. With globalization, the local is not so local anymore. Just as online commerce enables small businesses in middle America to reach global audiences, so too has the penetration of communications technologies in Colombia, Somalia, and Uzbekistan given those geographic outposts global reach. For new digital natives cyberspace may offer not only the best means for routing around structural barriers to socioeconomic advancement, but also a chance to access global markets and economic riches far in excess of those available locally. Such access does not require venture capital, leased office space, and a large staff; it requires intelligence, boldness, and Internet connectivity through a cheap consumer device.

As these next billion digital natives come online, Western assumptions about cyberspace will inevitably be challenged. Communications technologies are neither empty vessels nor forces unto themselves. Rather, they are complicated, continuously evolving manifestations of social forces at a particular time and place. Once created, communications technologies in turn shape and limit the prospects for human interaction in a constantly iterative manner. It's hard to say what cyberspace will look like twenty years from now, but one thing is certain: we won't be in Kansas anymore.

6.

We the People of . . . Facebook

In June 2012, Google's ongoing acrimonious relationship with China took a new turn. In order to assist users to access information freely from behind the Great Firewall of China, Google created a unique feature on its search engine: when users search for banned keywords, it warns them that their connection will not work and suggests alternative spellings and phrasings that will take them to the same content. In effect, Google's search engine now facilitates the circumvention of China's censorship of the Internet.

A second new Google feature, introduced in the same month, sounds a more ominous tone: "WARNING: We believe state-sponsored attackers may be attempting to compromise your account or computer. Protect yourself now." Google is tight-lipped about how it determines that a particular attack is state-sponsored, and does not explain to what extent the company would issue this warning to users who might be victimized by states with which Google does not have an antagonistic relationship. In June 2012 the *New York Times* reported that the U.S. and Israel were behind the Stuxnet virus that sabotaged nuclear enrichment facilities in Iran. Would Google give the same warning to Iranians working in critical infrastructure facilities?

The new Google features certainly irritated the Chinese leadership, and the company demonstrated a degree of boldness that few others are willing to embrace, especially in the scramble to get into the Chinese market. But the issues at play have implications far beyond the ongoing tensions between Google and China. They are part of a larger trend emerging in cyberspace: the growing political importance of the corporate giants who own and operate the technological domain. The decisions they make for commercial reasons often have political consequences, both domestically and internationally. The companies that control huge swaths of cyber-space are at once flexing their political muscles and being depu-tized with more expansive policing responsibilities. Are we entering a new age of corporate power in cyberspace? Are companies like Facebook and Google examples of a new type of "corporate sover-eignty" as Rebecca MacKinnon, author of *Consent of the Networked: The Worldwide Struggle for Internet Freedom*, suggests?

• • •

The world has come increasingly to depend on a globally networked information and communications infrastructure, and data on our desktops is now entrusted to servers and networks beyond our control, distributed across territorial jurisdictions and around the planet. The infrastructure for this complex network is primarily owned and operated by the private sector: companies that run our Internet services, telecommunications networks, mobile phones, and satellites. To that market, we have added another: the personal data we give away for free. All of our "likes," "pokes," and "tweets" are geolocated and cross-referenced with our purchasing habits, social networks, and professional interests, and sold to com-panies that can then more accurately target their advertisements. As more and more profit is to be made from exploiting big data, the urgency to acquire, store, mine, and analyze more and more person-alized information grows. It is a self-reinforcing business model that extends the reach of companies ever deeper into our personal lives and, in turn, increases our own dependence on the platforms they provide and control. We can no longer function properly without access to our email accounts, cellphones, and cloud-based data not because we are addicted to them per se (although many of us are) but because we *need* them to live and work.

The companies behind all of this have become behemoths. In 2012, Facebook announced that it had over 1 billion subscribers, more than the entire Internet-using population in 2005. If it were a country, this would make it the third largest on earth, after China and India. Notwithstanding that its IPO failed to meet expectations, Facebook's 2012 market capitalization was more than $60 billion. But Facebook is puny compared with some of its competitors: Apple's quarterly profits in 2012 exceeded $11 billion, and its market capitalization was roughly $525 billion, more than Microsoft's ($244 billion) and Google's ($192 billion) combined.

The sheer size of these companies, combined with our dependence on them for all of our communications needs (and experiences), makes their decisions and policies enormously consequential. As MacKinnon argues in *Consent of the Networked*: "We have a problem: the political discourse in the U.S. and in many other democracies now depends increasingly on privately owned and operated digital intermediaries. Whether unpopular, controversial, and contested speech has the right to exist on these platforms is left up to unelected corporate executives, who are under no legal obligation to justify their decisions."

As these companies grow and mature, we should expect them to exercise even more political influence both at home and abroad. Google's lobbying and advocacy activities are a case in point of a company attempting to shape public policy in accordance with its commercial interests. In addition to issuing user warnings, Google has supported free-speech activists and research networks, and in

2011–2012 held two major "Internet at Liberty" conferences designed to raise awareness about threats to an open Internet and provide opportunities for free-speech activists to network and share ideas. Its vigorous opposition to the SOPA and PROTECT IP (PIPA) bills – the Stop Online Piracy Act and the Preventing Real Online Threats to Economic Creativity and Theft of Intellectual Property Act – which together aim to curtail online copyright violations by granting the U.S. government new tools and powers to block user access to websites that sell copyright-infringing or counterfeit goods, were seen by many as instrumental to the bills' defeat in the U.S. Congress.

Of course, having Google lobby on behalf of Internet openness is welcome, but there are justifiable concerns about the implications of such a wealthy and powerful company throwing its weight behind political causes in a selective and partisan manner. What if activists turn their attention to Google itself? If Google funds them, will they temper their criticism? What about Google's resistance to privacy protections? Is its support for human rights online selectively applied to those areas that mesh with the company's private interests? (A full disclosure is in order: like many other Internet research groups, the Citizen Lab has benefited from Google donations and, particularly, it is a host organization in the annual Google Policy Fellow Program, which funds a visiting researcher placed at the Citizen Lab each summer. This support is fully transparent and comes with no strings attached. Should that change, the Citizen Lab would not accept financial support from Google.)

We have come to depend on social media like Twitter and Facebook as the online equivalent of the public square, and we use these platforms not just for entertainment but also for political discourse. The companies themselves sometimes contribute. Beginning in 2009, for example, Facebook offered its users the opportunity to vote on its privacy policy, making its some billion users the largest electorate in the world. But initiatives in this direction can also have their limits: In 2012 Facebook hinted that this voting structure was becoming burdensome, even unworkable. The company posted a statement saying it "found that the voting mechanism, which is triggered by a specific number of comments, actually resulted in a system that incentivized the quantity of comments over their quality. Therefore, we're proposing to end the voting component of the process in favor of a system that leads to more meaningful feedback and engagement." So much for direct corporate democracy.

Rarely do average citizens step back and examine the constraints and opportunities presented by these companies, the "unprecedented synthesis of corporate and public spaces," as Steve Coll put it in his 2012 *New Yorker* essay, "Leaving Facebookistan." As the Electronic Frontier Foundation's Jillian York has suggested, social media are less like town squares and more like shopping malls, bound by private regulations. Decisions taken by these companies for commercial reasons end up having enormous consequences for freedom of speech, association, and access to information. It is important to remind ourselves of the political economy of social media: If social media seems like "imagined communities," in the language of Cornell University professor and author Benedict Anderson, the members are more like serfs than citizens, the users both consumers and product. Social media might thus best be described as *epiphenomenal* public spheres: while we may increasingly use these platforms for political purposes, politics is only a by-product of their intended purpose, and one that is highly constrained by terms of service that are outside the direct control of users.

● ● ●

The epiphenomenal nature of political participation in social media is exacerbated when one factors territorial jurisdiction into the equation. While almost all social media platforms have international customer bases, they are registered and headquartered in particular political jurisdictions, and are subject to the laws and regulations of those jurisdictions. As mentioned earlier, any data stored on Google servers, no matter their physical location, is subject to Patriot Act provisions on data sharing because Google is domiciled in the United States. (In 2012, in response to this reality, Norwegian legislators proposed regulations that would restrict the use of Google products and services by public officials.) More generally, when we use Gmail, Facebook, and other social media platforms, we may be subjecting personal data to laws and regulations over which we have no direct control.

While social media companies may wield growing political power, in turn they are increasingly subject to the assertion of state power in cyberspace, particularly around security concerns. There has been a sea change over the past decade in the way governments approach cyberspace security and governance. Whereas in the early days of the Internet, state policy was either absent or deliberately hands-off, today governments seek to shape and secure cyberspace as an urgent priority. Because much of what constitutes this realm is in private sector hands, in order to secure cyberspace governments must enlist or otherwise compel the private sector to police the data and networks they control within their territorial jurisdictions. These pressures have led to a gradual downloading of policing responsibilities to the private sector to monitor users, filter access to information, and control free speech.

How and under what authority the private sector polices cyberspace varies widely between political jurisdictions. Compliance with local laws often brings with it tough choices that, if taken, compromise larger principles, especially those related to human rights and privacy protections. Companies like Google, Microsoft, Research in Motion, Yahoo!, Twitter, Facebook, and many others have all faced growing pressures as their operations have expanded worldwide, and they have had to balance the desire to penetrate markets and the need to comply with local laws to do so against respect for freedom of speech and access to information. In September 2012, for example, a Brazilian court ordered the arrest of a Google executive when the company refused to remove an online video that criticized a mayoral candidate in the country's upcoming municipal elections. The charge was that the video violated a local pre-election law which prohibits "offending the dignity or decorum of a candidate." In a statement to the *Washington Post*, Google explained that it "is appealing the decision that ordered the removal of the YouTube video because, in being a platform, Google is not responsible for the content posted on its site." Earlier that month, a different Brazilian judge fined Google for another video that criticized another candidate, and yet another Brazilian judge ordered the arrest of another Google official, a decision that ended up being overruled by a higher court. Needless to say, Google and Brazil have issues, and Brazil is not alone. In February 2010, the Italian government charged four Google executives over the posting of a video of a bullying incident involving an autistic boy in 2006 that a court ruled violated privacy laws. Three of the four executives were handed six-month suspended sentences.

And then consider the case of India, another democratic country that has pushed for an increasingly stringent set of requirements on Internet, social media, and mobile providers to police cyberspace through its 2011 Internet Intermediaries Guidelines. In 2012, the Indian government demanded that Yahoo!, Gmail (owned by Google), and other email providers store all emails accessed in India on Indian-based servers, even if the mail account was registered outside the country. After a prolonged and largely secret negotiation, RIM was compelled to do likewise for its popular BlackBerry service.

The same downloading of responsibilities can be seen in legislative proposals dealing with copyright enforcement. The Anti-Counterfeiting Trade Agreement (ACTA), signed by the United States, the European Union, Mexico, South Korea, Singapore, New Zealand, Australia, and Canada (and ratified by Japan) is hugely controversial owing to (among other reasons) its broad definition of criminal liability, which would hold the private sector legally responsible for what users do or share through their services. How such downloading of controls might be regressive is illustrated in the case of Twitter's new "micro-censorship" policy, engineered to remove tweets only of Twitter users based in countries whose governments have asked Twitter to remove content. Although the company assuaged many critics by being transparent about the new policy and stating that it archives content removal requests with the Chilling Effects Clearinghouse, a 2012 incident in France showed how such a policy can go terribly wrong in practice. Four accounts that parodied and impersonated the French president were terminated by Twitter, ostensibly because they violated Twitter's terms of use policies around parodying. Although many citizens and observers disagreed with this assessment and were outraged, nothing was done to reverse the policy. When private companies are entrusted with the responsibilities and powers to police the Internet, questions of transparency, due process, and accountability inevitably arise.

• • •

The cyber behemoths of the social media world may be formidable, but they are hardly the first examples of large corporations wielding political influence. Going back in time, multinational corporations like England's Hudson's Bay Company and East India Company effectively ruled large parts of Canada and India in ways that Google, Facebook, and others could never hope to match in scope and scale. Throughout the twentieth century, political economists lamented the rise and influence of multinational corporations who used their power and influence to manipulate the domestic political systems of the countries within which they operated. Similarly, resource extraction giants throw their political weight around in Washington, with a degree of sophistication that makes the Googles and Facebooks of the world look like amateurs. The relationship between the private sector, public authority, and citizen and consumer rights is constantly shifting, part of the ongoing process of industrialization, globalization and, with luck, increased democracy worldwide. Social media giants are but the latest manifestation of this dynamic.

But there is a difference. Social media giants may be just the latest incarnation of age-old corporate behemoths, but the markets they control are qualitatively different. As Rebecca MacKinnon says, "Not only do they create and sell products, they also provide and shape the digital spaces upon which citizens increasingly depend." Those digital spaces include not only the spheres through which the public interacts but also increasingly a large amount of the space where we conduct our private lives – our close social relationships, even our innermost thoughts. As they grow in scope and scale, and take on a more important role in society, will citizens increasingly confuse social media companies with political institutions? Will political authorities continue to delegate core regulatory responsibilities to the private sector?

7.

Policing Cyberspace :
Is There an "Other Request" on the Line?

> No free man shall be seized or imprisoned, or stripped of
> his rights or possessions, or outlawed or exiled, or deprived
> of his standing in any other way, nor will we proceed with
> force against him, or send others to do so, except by the
> lawful judgment of his equals or by the law of the land.
>
> — Clause 39, Magna Carta, 1215 CE

In November 2012, Google released an update to its bi-annual "Transparency Report." The reports cover six months of traffic to Google services from around the world (in this case, from January to June 2012), information essential for researchers attempting to verify blockages coming from certain countries that indicate tampering with Internet connectivity. For instance, this data helped illustrate the Internet blockades that occurred in Egypt and Libya during the 2011 Arab Spring. The reports also disclose the number of "removal requests" Google receives from copyright owners and governments, and the number of "user data requests" it receives from government agencies and courts. Until 2012 such disclosures were unique, but a week after Google's June update, Twitter released a transparency report of its own.

Does this represent a growing trend towards corporate transparency? Google's transparency efforts have their roots in the company's contentious attempt to penetrate the Chinese market, which had led to numerous requests from the Chinese government for blocks and for the release of user data and other information controlled by the company. But the issue goes far beyond China. Google (and other Internet companies) now routinely face a barrage of requests from governments worldwide looking to secure cyberspace, and to exert control over the companies that own and operate it. The Google and Twitter transparency reports are but a symptom of mounting government pressures. According to Google's November 2012 report, "The number of government requests to remove content from our services was largely flat from 2009 to 2011. But it spiked in this reporting period. In the first half of 2012, there were 1,791 requests from government officials around the world to remove 17,746 pieces of content."

While Google does not disclose the details of every request, it does highlight those that it complied with or turned down. Examples of requests rejected by Google include: Passport Canada's plea to remove a YouTube video of a Canadian citizen urinating on his passport and flushing it down the toilet; Pakistan Ministry of Information's request to remove six YouTube videos that satirized the Pakistani Army and senior politicians; the Polish Agency for Enterprise Development's request that Google remove a search result that criticized the agency, as well as eight more that linked to it; and the Spanish Data Protection Authority's plea that Google remove 270 search results that linked to blogs and sites referencing individuals and public figures, plus several videos hosted on YouTube that did the same.

• • •

There are "requests" and then there are "other requests"; that is, there are legitimate reasons to shut down certain websites or remove certain content and then there are murky, theatre of the absurd, non-court-ordered "other requests" that governments the world over are increasingly making. These "other requests" tread dangerously close to our "other lives" – the secrets we share, for very good reason, only with a select few intimate "others" – and our lives, our sanity, depend on the ability to do so, to remain anonymous, invisible to the world outside, particularly to those in positions of power.

Google arranges its content removal request results in two categories: those that come with a court order and those that do not. The only information Google provides in elaborating on these "other requests" is contained in brackets – from "Executive," the "Police," et cetera – and such requests often dwarf the number of requests that come with a court order. For example, in 2011 Google received five requests for content removal from India that came with a court order, but nearly 100 for the removal of 246 items that came in the form of "other requests," of which Google complied with roughly one-quarter. The request from Passport Canada was an "other request," one of fifteen such requests from Canada, as opposed to four that came backed with a court order. In total, "other requests" to Google increased from 563 in the July to December 2011 period to 1,002 from January to June 2012.

In 2009 a senior Google staffer who dealt regularly with China told me in confidence about the persistent number of "other requests" Chinese authorities make to Google. While many experts assumed that Chinese officials were mostly concerned with data about the Falun Gong, Tiananmen Square, Tibet, or Taiwan, the vast majority of requests were far more frivolous: senior government officials wanting an embarrassing YouTube video of their daughters taken down, or evidence of bureaucratic inefficiency erased from the public record. Google staffers had no policy to deal with such requests and were often at the whim of capricious officials. Most frustrating, I was told, was that they would be given little time to deal with such requests. Failing to comply, often within less than twenty-four hours, could lead to stalling around approvals necessary to run their business. More than the 2009 Operation Aurora attacks, in which Gmail accounts and Google source code were infiltrated by Chinese hackers, it was the frustration of operating in this "other request" climate that ultimately led to the company's withdrawal from mainland China to Hong Kong, and to initiatives like the Transparency Reports.

The number of "other requests" Google receives for content removal or user data is suggestive of a troubling trend. Looking at their Transparency Reports, and the "other requests" category in particular, it is striking how much of the policing of cyberspace is taking place outside of the rule of law. As more and more data is entrusted by users to third parties like Google, governments are side-stepping transparent and accountable judicial processes to police that data. "Other requests" are, ironically, becoming the norm for Internet policing.

The most obvious place we would expect to find evidence of this trend is in flawed democracies or autocratic regimes. In Google's 2011 (July–December) report, Turkey made twenty-three requests for seventy items to be removed, of which 48 percent were complied with; Taiwan, eight content removal requests for twenty-seven items, none of which were complied with; Pakistan, two requests to remove fifteen items, neither of which was complied with. But the most shocking revelation of the Transparency Report concerns those "other requests" for user data and content removal from liberal democratic countries. According to Google, between January and June 2012, the ten countries making the most "other requests" are all democracies: Turkey, United Kingdom, Germany, India, United States, Spain, Brazil, France, South Korea, and Canada. Is it a crime to urinate on your passport in Canada? I doubt it. But it most definitely is wrong for the Canadian government to make a request to an ISP to remove a video documenting the act. If it were a criminal offence to urinate on your country's passport, arguably the video is material evidence and, more importantly, what should or should not be censored in Canada is not a decision Passport Canada is authorized to make.

While the Canadian example stands out for its arbitrariness, the data concerning the United States stands out for sheer volume. According to Google, between July and December 2011, seventy "other requests" were made by the U.S. government for the removal of 2,341 items, of which 44 percent were complied with. In more than half of the cases, Google determined these "other requests" to be frivolous. Germany made forty-three "other requests" for the removal of 418 items, of which 72 percent were complied with. In Britain, there were thirty-seven "other requests" for the removal of 750 items, of which 54 percent were complied with. France made nineteen "other requests" for the removal of thirty-nine items, of which 47 percent were complied with. With the exception of some minor notes made by Google in the margins, no further detail is provided. At least in this instance, we are left largely in the dark about how cyberspace is being policed.

A short time after Google's first 2012 update, Twitter produced its own transparency report. The company noted its commitment to responsible behaviour and the inspiration provided by Google. Perhaps more tellingly, however, Twitter's report came out immediately after it lost a court case and was required to turn over three months' worth of data on an Occupy Wall Street protester. Although their transparency report does not provide anywhere near the same level of detail as Google's, it is an interesting complement and represents the same troubling trend. According to Twitter, the company received more requests from governments in the first half of 2012 than in all of 2011. When it comes to requests for user information, no country comes close to the United States: 849 user information requests, and of those Twitter complied with 75 percent. For its part, Canada made eleven requests for user account information and Twitter complied with two. Japan made ninety-eight requests, nineteen of which were complied with. Twitter is one of the few companies that alerts users to requests made by law enforcement agencies about their accounts, unless specifically prohibited from doing so by a court order or statute.

We might assume that it is the countries of the global South and East that are the main threats to rights and freedoms in cyberspace, but the global North and West appear to be on the same path – maybe even leading the way.

• • •

As far as transparency practices go, the Google and Twitter reports are a step in the right direction, but they represent only a partial peek into the hidden underworld of extrajudicial cyberspace policing. Also, the reports beg several questions. What other types of requests are being made that are not anywhere disclosed? To what extent do other companies receive the same type and volume of requests? Presumably, Google and Twitter are representative of a much larger social media universe. What about Microsoft? Skype? Facebook?

The Electronic Frontier Foundation (EFF) has attempted to answer these questions on a project website called "When the Government Comes Knocking, Who Has Your Back?" The EFF has investigated and ranked eighteen U.S. email, ISP, and cloud storage companies across several categories – including terms of service, privacy policies, and published law enforcement guidelines – and examined company track records of standing up for their users in court. As the EFF explains: "When you use the Internet, you entrust your online conversations, thoughts, experiences, locations, photos, and more to companies like Google, AT&T and Facebook. But what happens when the government demands that these companies hand over your private information? Will the company stand with you? Will it tell you that the government is looking for your data so that you can take steps to protect yourself?" Their scorecard is instructive: not one of Amazon, Apple, AT&T, Comcast, Foursquare, Loopt, Microsoft, MySpace, Skype, Verizon, or Yahoo! tells users about data demands or are transparent about government requests.

While the Google and Twitter transparency reports, and projects like EFF's, provide some insight, other research fills in the blanks and helps illustrate the danger of "other requests" becoming the norm vis-à-vis cyberspace. Take the case of Anthony Chai. In 2006, this Thai-American citizen was detained and interrogated while visiting Thailand, and subsequently harassed when he returned to the U.S., for comments he'd posted on http://www.manusaya.com a year earlier that allegedly violated Thailand's *lèse-majesté* laws (which make it a crime to insult Thai royalty). The website was hosted by Netfirms, a Canadian company, which shut it down in June 2005 at the Thai government's request. Because Netfirms is not a Thai company and enjoyed the protections of Canadian law, the shutdown was an especially pernicious response to an "other request," – one with considerable international implications. Equally disturbing is the possibility that Netfirms released Anthony Chai's IP address, linking anonymous http://www.manusaya.com posts to him and subjecting him to detention and harassment by the Thai government. This disclosure would be in violation of Canada's Personal Information Protection and Electronic Documents Act (PIPEDA). Though the case is pending in a California district court (where Chai has filed a lawsuit against Netfirms), it shows that a company may violate its own country's laws in seeking to comply with another country's "other requests."

One country where "other requests" are the operational norm is China, and probably the most infamous case there involved Google's competitor, Yahoo! In 2002 and 2004, the Chinese government requested information – including private emails – about two dissidents it sought to silence and punish: Wang Xiaoning and Shi Tao. Yahoo! complied with the demands and the sharing of their data led directly to the prosecution and imprisonment of the dissidents. Their families, with the help of the American branch of the World Organization for Human Rights, filed a suit against Yahoo! in the United States, and Yahoo! executives were called before Congress to explain their actions. The company, they testified, was merely following "local law." Compliance with the type of demands that Yahoo! faced in China might well be part of doing business in that country, but where should foreign companies draw the line?

In 2008, as previously mentioned, the Citizen Lab's Nart Villeneuve discovered that there was a content filtering system on the Chinese version of Skype. Called "TOM-Skype," it is a joint venture between Skype (which at the time was owned by eBay, but is now owned by Microsoft) and the Chinese media conglomerate, the TOM Group. The Citizen Lab looked into TOM-Skype's content filtering mechanism, and found that each time certain keywords were typed into the chat window a hidden connection was made. We followed that hidden electronic trail, apparent only through detailed packet capture analysis, to a server in China which, it turned out, had a directory that was not password protected. It contained a voluminous number of encrypted files, plus the decryption key. Upon decrypting the data, we discovered that TOM-Skype had been systematically intercepting and monitoring millions of private chats, triggered whenever any of the users typed in a banned keyword. From that moment on, their chats were intercepted, as were those with whom they were communicating, and uploaded to a server in China, presumably to be shared with

Chinese security services. The interception directly contravened Skype's explicit terms of service, which promised state-of-the-art "end-to-end encryption," allowing it to be widely promoted as a secure tool for dissidents and others at risk.

The scandalous tale was covered by John Markoff in the *New York Times,* and Skype later apologized. A few years later, however, University of New Mexico researchers found the exact same content-filtering and interception system was still in place on TOM-Skype. Notably, Skype scores zero on the EFF scorecard, and its present owner, Microsoft, fares little better: neither tells users about data demands, is transparent about government requests, or fights for user privacy rights in court. Apparently, "local laws" and "other requests" prevail.

• • •

In the case of WikiLeaks, although no judicial process supported it, many companies either pulled their services or refused to support the organization after it linked to thousands of leaked U.S. State Department cables. In December 2010, its domain name service provider, EveryDNS, ceased DNS-resolution services for http://www.wikileaks.org, severely hampering its ability to communicate. EveryDNS cited the ongoing denial-of-service attacks against WikiLeaks as the reason for its cessation of services, but most suspect the U.S. company was wary of political repercussions in the event of continued service. Another American hosting company, Amazon, also dropped WikiLeaks as a customer. And, around the same time, several credit card and financial services companies – Bank of America, Visa, Western Union, MasterCard, PayPal, and Amazon – stopped processing donations to WikiLeaks. PayPal claimed it did so because WikiLeaks violated its terms of service, which states, "Our payment service cannot be used for any activities that encourage, promote, facilitate or instruct others to engage in illegal activity." One PayPal executive, Osama Bedier, claimed the company took the measure after a letter was circulated by the State Department that referred to WikiLeaks being in illegal possession of documents. However, nowhere does the letter, sent directly to WikiLeaks, suggest the organization itself was breaking U.S. law, and this raises the troubling prospect of a government and/or company arbitrarily deciding to withhold services to an organization simply because it is controversial.

Not all "other requests" to police the Internet come from government agencies: many come from the private sector, typically under the rubric of intellectual property enforcement. When Cryptome.org published a leaked version of Microsoft's global law enforcement guide, Microsoft sent a Digital Millennium Copyright Act (DMCA) take-down notice to Network Solutions, the DNS and hosting provider for Cryptome.org. Network Solutions' response was to shut down the entire Cryptome website, which had on it thousands of leaked sensitive national security documents having nothing to do with the Microsoft case. In other words, to deal with one potentially diseased apple, Network Solutions decided to raze the entire orchard. As *Wired* magazine's "Threat Level" put it, "Microsoft has managed to do what a roomful of secretive, three-letter government agencies have wanted to do for years: get the whistle-blowing, government-document sharing site Cryptome shut down." Although the site was eventually restored, the matter illustrates just how "other requests" can ripple outwards like a percussive wave in the wake of a major blast.

Some instances of "other requests" coming from private actors are used to exploit companies' terms of service and their willingness to err on the safe side of controversy. In July 2012, Facebook erased a status update of free speech organization Article 19. Article 19's post linked to a series of satellite images from a Human Rights

Watch report outlining where it suspected Syrian security services were engaging in torture. Without providing any notice, Facebook erased the post. It later said the post was mistakenly removed by a member of its "moderation team" after it received a terms of service violation complaint. Article 19 did not see it that way. Dr. Agnès Callamard, Article 19's executive director, said, "The deletion shows the looming threat of private censorship. We commend Facebook for creating tools to report abuse, but if your post was wrongly deleted for any reason, there is no way to appeal. Facebook doesn't notify you before deleting a comment and they don't tell you why after they have. Facebook acts like judge, jury, and executioner."

• • •

"Other requests" can land companies between a rock and a hard place: the growing demands by governments, often delivered with a hint of serious consequences for noncompliance, and the prospect of outrage and controversy from an aggrieved public demanding the "right to know" make it a perilous situation. In 2010, Research in Motion (now known just as BlackBerry), the Canadian-based manufacturer of BlackBerry devices, faced numerous demands by governments to eavesdrop on users of its products and services. The demands started with Middle East countries, including the United Arab Emirates and Saudi Arabia, then spread to Indonesia, India, and others. Each country insisted that RIM give government agencies backdoor access to its encrypted data streams, something RIM claimed was contrary to the technical design of its infrastructure, and thus impossible. Governments, observers, and analysts were confused by the issue, a state of affairs not alleviated by RIM.

On the one hand, the company claimed that its services were so secure that even RIM itself could not decrypt its own encrypted data streams. "RIM would simply be unable to accommodate any request for a copy of a customer's encryption key since at no time does RIM, or any wireless network operator, ever possess a copy of the key," the company said in a statement. On the other hand, the company also said it respects "both the regulatory requirements of government and the security and privacy needs of corporations and consumers." But how are these two principles reconciled when governments require access to data for law enforcement and intelligence purposes? RIM considers its negotiations with governments about access to be "confidential," yet says it doesn't make special arrangements with one country that aren't "offered to the governments of all countries." If that's the case, why are there confidential negotiations at all? There has also been confusion about which of the many RIM services and products are secure, and which are not. RIM says "customers of the BlackBerry Enterprise Solution can maintain confidence in the integrity of the security architecture without fear of compromise." Does this mean its much more widely distributed consumer-level product, the basic BlackBerry, is less secure and can be easily monitored?

To help answer these questions, in 2011 the Citizen Lab set up a publicly announced project called RIM Check, a specially designed website in which users of RIM products were encouraged to fill out a series of questions about their usage. The website would collect the IP address and information about the device used, hopefully showing the route the request took based on the type of BlackBerry product being used. Our theory was that if RIM made arrangements with certain countries the exit point from the RIM network might show up on a server in a particular jurisdiction where it should not be. We also monitored for content filtering over the BlackBerry network.

Although rarely mentioned at the time, the RIM controversy went beyond the interception of data. A BlackBerry is also used to surf the Web, and in many of the countries where RIM was being pressured Internet filtering is de rigueur. A Kuwaiti newspaper reported that RIM agreed to filter access to 3,000 pornographic websites at the regime's request, and some users reported to us that RIM was already filtering access to Web content in the U.A.E. and Pakistan. Preliminary tests done in Indonesia suggested it might be going on there too. Although the data from the RIM Check project was too unreliable to draw firm conclusions (and we never published a final report for that reason), it did raise critically important questions and considerable public awareness about the issue.

However much our RIM Check project was a thorn in the company's side, it must have been only a minor irritation compared with the real deal: the demands being made on RIM by governments for access to its encrypted data streams were jeopardizing the company's "secret sauce," calling into question one of its most marketable components, its supposedly "unbreakable" communications network. Unfortunately, RIM's strategy consisted mostly of saying as little as possible in the hope that the controversy would magically disappear, and its attitude about the issue was plainly visible when co-founder Mike Lazaridis petulantly terminated an interview with the BBC after being asked whether the company had secretly made arrangements to share its data with governments in the Middle East, India, and elsewhere. "C'mon, this is a national security issue, turn that off," Lazaridis barked, ripping off his microphone and leaving his seat while the cameras rolled. Naturally, the video went viral.

• • •

Lazaridis's comments about the matter being a "national security" issue are telling. The trend towards "other requests" is part of the securitization of cyberspace, the slow transformation of an issue into a matter of national security, with new policies and controls attached. As the Internet has become an integral part of everyday life, how it is constituted and by whom has become a critical issue. Securitization opens the door to clandestine arrangements, over-classification, and lack of accountability. Often operating in the shadows and not subject to rules and regulations, those in favour of securitization insist that national security requires that governments have the freedom to manoeuvre and make rapid responses to immediate threats, even at the expense of cyberspace users' rights.

These shifting forces become more pronounced during major events. Terrorist attacks – like 9/11 or the London Underground bombings – are like political earthquakes that unsettle the existing system of checks and balances and trigger an avalanche of legislation that under normal conditions would seem excessive. A few short weeks after 9/11, governments around the world passed anti-terror legislation with most of the statutes featuring similar fundamental components: beefed-up domestic policing powers, relaxed restrictions on the sharing of information between domestic law enforcement and foreign intelligence services; new requirements on the private sector to retain and share with security services the data they control; and, most importantly, a loosening of the requirements for judicial oversight into matters of law enforcement and intelligence. A June 2012 Human Rights Watch (HRW) report found that 144 countries have passed anti-terror laws since September 11, 2001, most of them covering a wide range of activities far beyond what is generally understood as terrorism and allowing for much wider latitude and action on the part of law enforcement and intelligence agencies. As the report noted, when viewed as a whole these laws "represent a broad and dangerous expansion of government powers to investigate, arrest, detain, and prosecute individuals at the expense of due process, judicial oversight, and public transparency. Such laws merit close attention, not only because many of them restrict or violate the rights of suspects, but also because they can be and have been used to stifle peaceful political dissent or to target particular religious, ethnic, or social groups." We still live in the shadow of 9/11 as the endless war on terror proves, and an open cyberspace may become the ultimate victim.

Cyberspace securitization is reinforced by international co-operation: governments and industry leaders share best practices and information, and develop new laws based on mutual experiences. Such international co-operation can lead to greater openness and mutual transparency, but the opposite is just as likely: international institutions can become the loci for the imposition of illiberal policies and greater government control. As HRW found, one of the chief reasons so many countries adopted anti-terror legislation after 9/11 is that the UN Security Council passed several resolutions urging member states to do so. This led to what HRW calls a "flood of new and revised laws that granted special law-enforcement and other prosecutorial powers to the police and other authorities."

Security and surveillance practices are also reinforced by networks of telecommunications companies that work in collaboration with government agencies, share expertise, develop standards and solutions, and harmonize practices. One such network is the Alliance for Telecommunications Industry Solutions (ATIS), a North America–focused consortium with more than 180 members from law enforcement and industry, including Public Safety Canada, Department of National Defence (Canada), the FBI's Electronic Surveillance Technology Section, AT&T, Microsoft, Bell Canada, and Verizon. ATIS hosts a number of committees and sub-committees, some of which focus specifically on developing standards for lawful intercept, such as the cumbersomely titled Packet Technologies and Systems Committee's Lawfully Authorized Electronic Surveillance (LAES) subcommittee, currently working on standards for Voice over Internet Protocol (VOIP) services. ATIS has a counterpart in Europe called the European Telecommunications Standards Institute (ETSI), whose meetings are also attended by the world's largest telecommunications companies, and law enforcement and intelligence agencies such as Public Safety Canada and the United Kingdom's Government Communications Headquarters. Such regular meetings help explain how and why countries like the United States, Canada, Australia, and the United Kingdom are all tilting towards shared policies around surveillance practices. The inside-the-club nature of the meetings – journalists and regular citizens cannot apply for membership in ETSI – may help explain why they're also gravitating towards limiting the basic judicial protections at the core of liberal democracy.

The Council of Europe's Convention on Cybercrime – an international convention meant to coordinate law enforcement practices among member states – is another case in point, and dozens of governments party to this agreement (including many non-European states like Canada and the U.S.) are in the process of ratifying it through national legislatures. In Britain, the proposed Communications Data Bill – an update to the Regulation of Investigatory Powers Act (RIPA) – would require ISPs and other telecommunications companies to store a wider range of communication data (such as use of social networking sites, VOIP services, and email) accessed in near-real time by law enforcement without a warrant. Under the bill, ISPs would have to route data via a "black box" that will separate "content" from "header data" and also have the capability to decrypt encrypted communications (such as transmissions over encrypted SSL – Secure Sockets Layer – channels). The bill has been widely criticized across the private sector, civil society, and inside the government itself. Wikipedia's Jimmy Wales threatened to encrypt all communications to the U.K., and stated: "It is not the sort of thing I'd expect from a Western democracy. It is the kind of thing I would expect from the Iranians or the

Chinese." Dominic Raab, a Conservative MP, said: "The use of data mining and black boxes to monitor everyone's phone, email and web-based communications is a sobering thought that would give Britain the most intrusive surveillance regime in the West."

The necessity to conform to the Convention on Cybercrime has often been cited by Canada's Conservative government as the impetus behind "lawful access" bills in Canada, the latest manifestation of which was Bill C-30 – the so-called Protecting Children from Internet Predators Act. That bill was politically mishandled, with Public Safety Minister Vic Toews infamously declaring in Parliament that you are "either with us or with the child pornographers," leading to a major public backlash that included a prominent Twitter "#TellVicEverything" campaign in which users tweeted inane details of their daily lives. While the government shelved the bill, lawful access legislation will invariably return in another guise. The central components of the proposed legislation included expanding police powers, imposing equipment and training costs on telecoms and ISPs, enabling telecoms and ISPs to voluntarily provide consumer information to authorities without a warrant, forcing telecom companies and ISPs to provide detailed subscriber data without a warrant, and imposing gag orders on telecoms and ISPs that comply with lawful access powers. Taken together, it is as if the bill would legislate "other requests" as the domestic and international legal and operational norm.

While the bill's explicit details were ominous enough, a revelation that emerged almost by accident during the public debate was even more troubling. In making the case for the powers outlined in the proposed law, its backers accidentally let slip that Bill C-30 would legislate warrantless informal sharing of information *that was already going on between telecom companies and law enforcement and intelligence agencies*. From documents released under federal access to information laws, University of Victoria doctoral student and Internet privacy expert Christopher Parsons found that in 2010 the RCMP contacted ISPs for user name and address information more than 28,000 times without a warrant, with the ISPs complying nearly 95 percent of the time. Although meant to be a consolation to the bill's critics, the revelation instead confirmed their worst fears: Canadian telecommunications companies and ISPs were already sharing data with law enforcement and intelligence agencies outside of judicial review! The bill would simply legislate that existing practice into law. "Other requests," it seems, have been the norm in Canada for some time.

As I ponder these issues, I think of Public Safety Canada's "Building Resilience Against Terrorism, Canada's Counterterrorism Strategy," released in February 2012. This document warns of "extremism" and the possibility of "low-level violence by domestic issue-based groups" and appears to open the door to the surveillance of legitimate non-governmental advocacy organizations. The strategy also details the need to monitor "vulnerable individuals" who may be drawn into politically motivated violence. I wonder what such a policy could lead to when government agencies are empowered to access personal data from private companies. Where are the protections against abuse, the checks against politically motivated witch hunts?

The downloading of lawful access responsibilities to the private sector almost certainly will be reinforced by the opening up of new markets for the commercial exploitation of data. As companies are forced to surveil/police their networks and data, products and services are emerging that enable them to do so more effectively and efficiently. The American privacy researcher Chris Soghoian has studied how new policing responsibilities are affecting corporate behaviour, and how some companies derive revenues from charging fees for "lawful access." He notes that the volume of requests received by one U.S.-based wireless carrier,

Sprint, grew so large that its 110-member electronic surveillance team could not keep up. As a result, Sprint automated the process by developing an interface that gives government agents direct access to users' data: George Orwell's *1984* in one fell swoop. Of course, Sprint charges a fee for this access, a fee that law enforcement agencies from the police to the FBI are more than willing to pay. In 2011, the Sprint direct-access interface was used by law enforcement agents more than 8 million times!

• • •

While the securitization of cyberspace manifests itself in new laws and informal practices, part of the growing surge of "other requests" has to do with the incentives facing companies that own and operate cyberspace: when pressed with content take-down requests, the companies often opt for the cheap and easy solution rather than demanding due process, risking expensive legal battles, or getting expelled from lucrative markets. There are, of course, legitimate reasons for companies to comply with local laws and with law enforcement and intelligence agencies in the countries in which they operate. But increasingly such co-operation takes place in countries that do not have, or are watering down, legal checks and balances over cyber security. Also, many of these countries have a much broader notion of what constitutes a security threat, and too often human rights activists, political opposition groups, and free-speech advocates are included. In short, complying with "local law" can mean colluding with some very nasty regimes.

The trend towards "other requests" in cyberspace policing is a disturbing descent into the world of black code. We live in an era of unprecedented access to information, and many political parties campaign on platforms of transparency and openness. And yet, at the same time, we are gradually shifting the policing of cyberspace to a dark world largely free from public accountability and independent oversight. In entrusting more and more information to third parties, we are signing away legal protections that should be guaranteed by those who have access to our data. Perversely, in liberal democratic countries we are lowering the standards around basic rights to privacy just as the centre of cyberspace gravity is shifting to less democratic parts of the world.

Underpinning this state intrusion (with self-interested or directed corporate backing) is a bald-faced public relations campaign that says essentially this: "If you've got nothing to hide, you've got nothing to worry about." While it is abundantly clear that a generation raised on and through social media is extraordinarily lax about personal privacy; that, indeed, seems to make a point of "going public" as often as possible; that has shelved the secretly written diary stored in a personal lock-box for "Look at me!" Facebook exposure, it is also true that I don't have a single friend, or know a single person of any age, who doesn't have secrets that they want and need to keep to themselves. (Indeed, I wouldn't trust any other such person; it is what makes them individual and interesting.) And yet, this campaign has been and continues to be extraordinarily successful. Like bystanders refusing to get involved when they witness that a crime is afoot, we have collectively stood by as cyberspace has been, and continues to be, compromised.

Long ago, the Canadian prime minister Pierre Elliott Trudeau may have insisted that "the state has no business in the bedrooms of the nation," but the same does not hold true for privacy online. Today, our private chats are considered fair game, our need for online anonymity voided in the interests of "national security" and control over cyberspace. And so, we have the growing norm of "other requests," a phenomenon that clearly illustrates why it is so important to lift the lid on cyberspace, to ask who controls the domain and what are they doing with our data? What happens to our email after we hear the *woosh* of it leaving our screens? Is it shared with anyone without our consent? Under what circumstances? Many of the companies that own and operate the complex services and infrastructure of cyberspace reassure us that their services are secure and that our data remains confidential, but the devil is in the details, in those lengthy end-user licence agreements we agree to before using our BlackBerry, iPhone, or Gmail accounts. Take, for example, the Internet Services Privacy Policy of Rogers Yahoo! (my own ISP), which states the following: "Personal information collected for the Internet Service may be stored and processed in Canada, the United States or other countries and may be subject to the legal jurisdiction of these countries." Other countries? Really? My data can be processed in another country and subject to its laws? Which countries? Whose laws?

In the universe of "other requests," one can only guess.

8.

Meet Koobface: A Cyber Crime Snapshot

"I own them!" Nart Villeneuve said triumphantly.

"What do you mean, you own them?" I asked.

"Their entire database. The mother ship. Victims, referrals, revenues, cellphone numbers. Everything!"

"How?"

"Even bad guys have to back up their data."

There is a 1960s episode of *Star Trek* in which the main characters, Captain Kirk and Spock, are confronted by their evil doppelgängers, physically identical to them in every way. Fifty years later, Facebook, the world's largest social networking community, has been confronted by just such a doppelgänger: Koobface.

The Citizen Lab tracked the gang behind Koobface for months in 2010, watching their every move. Villeneuve was in it for the challenge of solving the puzzle, the thrill of the hunt. Our wider motivation, however, was to better understand cyber crime, and how crime, espionage, and warfare might be blurring together over and through cyberspace. Koobface was well known among the exclusive (and often contentious) club of technogeeks who study the malicious underworld of cyberspace. The Villeneuves of the world had been following it since the mid-2000s, when Koobface emerged as a menace to the growing social networking community that it so openly mocked and exploited. But no one had detailed knowledge of how Koobface worked or who the perpetrators behind its vast reach were. Was it a well-organized crime syndicate? A few bored teenagers? Something else, perhaps more nefarious? Whatever the truth, Koobface was trolling across the world of social networking like a giant digital amoeba, consuming and spitting out unsuspecting victims. In 2010, the Russian security company Kaspersky Lab estimated that Koobface controlled nearly 800,000 computers worldwide, each belonging to users lured into its trap.

Becoming ensnared happens easily. Koobface sends a link over Facebook from a "friend" (who has already been infected) that says something outrageous or provocative, something like "OMG! Have you see this naked video of you?" Who wouldn't follow that link? Or maybe another funny video of dancing kittens. Who wouldn't enjoy such a thing? But for the hapless recipient, that one curious click leads into an abyss of viruses and trojan horses, and straight into Koobface's grasp.

Koobface makes its money through pay-per-click and pay-per-install schemes. ("Pay-per-click" refers to a model whereby webmasters display third-party advertisements on their websites and earn income whenever Internet users click on these advertisement links. "Pay-per-install" refers to a model whereby the software of one company is promoted by a third party who is paid every time a user installs the software. Both are legitimate, but they have also been widely exploited by online criminals.) Once its malicious software is installed on a user's computer their Internet requests and website visits are redirected without their consent or knowledge to sites that pay Koobface for each visit. Some of these websites are themselves honeypots for yet more cyber crime, such as for phony antivirus software that promises to clean up your computer's hard drive by eliminating viruses and malignant files. Those initially victimized by Koobface are thus served up to other criminal entrepreneurs posing as good guys who promise to fix computer problems, but who, in fact, only make them worse. A cut of the revenue from the sale of fake antivirus products is then given to Koobface. Joint ventures, strategic alliances, globally distributed production chains, "value-added" services, robust customer management databases, and multiple (and complementary) revenue streams: Koobface is one sophisticated post-industrial operation.

• • •

Just the same, even the most meticulous criminals generally make mistakes. Our investigation really started with the discovery that Koobface backed up its entire database each and every day on a "zipped," or compressed, file, and that they did so on an Internet-connected computer without any password protection. It was left wide open, there for the taking, a mistake that laid bare the entire operation from the inside out.

Downloading and opening the compressed file gave us almost complete access to Koobface's operating infrastructure: how it worked (down to the finest detail), where the fraud was occurring, the worldwide locations of the compromised computers they had commandeered, and their revenue streams. We felt like voyeurs peeking through a window – with Koobface having no idea we were watching – justified in doing so given the lawless intent of those under our surveillance. Such privileged access gave us insight into the complexity and ingenuity of one of the world's leading cyber-crime outfits, and a richer understanding of the hidden swamps of cyberspace.

A major hurdle Koobface had to overcome was the precautions Facebook had in place to prevent fake "friends" using their trusted network. Each new Facebook account requires a real person to fill out a "CAPTCHA" – clusters of wavy, sometimes illegible letters and numbers. (CAPTCHA is an acronym for Completely Automated Public Turing Test to Tell Computers and Humans Apart.) As a standard security precaution, Facebook requires a human being to visually identify the CAPTCHAs and reproduce them in a field in order to create a new account.

To get around the CAPTCHA problem, Koobface engaged in what the cyber crime expert Marc Goodman calls "crime-sourcing," the outsourcing of all or part of a criminal act to a crowd of witting and unwitting individuals. With thousands of infected computers at its disposal, Koobface created a transnational assembly line of co-opted workers – the hapless computer owners themselves – who manually filled out the CAPTCHAs. It engineered a system through which a fake emergency pop-up window appeared on the screens of users throughout the world along with a box with the familiar Windows brand name and colour scheme carrying a startling warning: "Type the characters you see in the picture below. Time before shutdown 02:29, 02:28, 02:27, 02:26 . . ."

Faced with such a panic-inducing moment, most users complied simply to avoid the risk of having their computers crash and their work destroyed. And so, every day, by the thousands, in commandeered computers as far afield as Thailand, Canada, Mexico, China, and India, fake CAPTCHAs were entered, information fed through the Internet to the Koobface database, and from there to the legitimate Facebook account creation field, all properly sorted and organized in real time, with the account management system maintained by Koobface engineers. Problem solved.

Once the fraudulent Facebook accounts were created, Koobface encountered another hurdle: how would the enterprise accumulate "friends," the necessary conduit for revenue? Only the most careless people would accept a friend invitation on Facebook from just anyone, so Koobface created a system that automatically culled through and recycled accounts they had compromised, taking bits and pieces of people's identities to create Frankenstein-like friends. One person's images would be combined with another's birthday and status information, and these were combined with the "likes" and "dislikes," places of birth, and employment histories of other people. Combined in this way, that friend request from a vaguely familiar person might just be someone you knew from high school. *The name seems a little off, but I recognize that face from somewhere . . . Sure, why not? Accept.*

Shortly thereafter would come the enticing link with the naked video, the prompts to download viruses disguised as antivirus tools, and the emergency pop-up screens. A globally distributed, malicious organism feeding continuously off of our digital habits.

It was an ingenious scheme, and to keep track of revenues the Koobface group sent themselves text messages to their mobile phones summing up their daily spoils. As with other aspects of the operation, the organization here was meticulous. Intricate ledgers on every payment made and income received were kept. During the window we had on their system, a glimpse that lasted just less than a year, this part of the Koobface operation netted over $2 million. No doubt there were other revenue streams invisible to us.

We soon discovered that we were not the only ones to have access to Koobface's inner workings. Others, most notably the German cyber security researcher Jan Droemer, were on the same path. Droemer contacted us and we shared information and methods. While we were mostly interested in the morphology of Koobface, Droemer was more interested in "whodunnit." He had combined the same pieces of evidence – basically a broad sweep of open-source information: the website registration, forum postings made with nicknames associated with members of the group, coincidental findings of names, addresses, and phone numbers that both of us were able to cross-reference. In one spectacular piece of gumshoe work, Droemer discovered that photographs stored on the Koobface command-and-control machines contained metadata that pinpointed the geographic location of the gang right down to its St. Petersburg, Russia, headquarters. There, five Russian men between the ages of twenty and forty-five, decked out in Nike running shoes and polyester athletic gear and surrounded by iPhones and PowerBooks, led a casual work life straight out of a Silicon Valley startup. The Koobface "gang" turned out to be a group of guys in track pants living a very comfortable life in distant Russia, driving BMWs, and playing World of Warcraft while reaping millions of dollars a year.

• • •

As we prepared our report for publication, we debated whether and how to proceed with notification to proper authorities. Clearly a major global criminal operation was unfolding in real time before our eyes, but whom should we notify? Ever since *Tracking GhostNet* and *Shadows in the Cloud* were published, there were grumblings in Ottawa, questions about our methods and intentions. For our part, the internal operations of law enforcement agencies in Ottawa were a bit of a mystery. With the GhostNet investigation, for example, we turned over data to Public Safety Canada's Cyber Incident Response Centre in the hope that they would help notify victims. We never heard back from them. The Koobface investigation presented an opportunity for us to engage with Canadian law enforcement agencies once again, and hopefully assuage their concerns about the Citizen Lab. We also wanted to learn more about how law enforcement would deal with the evidence we had in hand.

We invited members of the RCMP's Integrated Technological Crime Unit to the University of Toronto and briefed them fully, turning over copies of the Koobface backups and walking them through the research we had done. The officers were grateful for the information, but seemed demoralized and fatalistic, intimating on several occasions that it was pointless for them even to begin an investigation. (One officer warned us against outing the group. "These might be the type of people who'll firebomb the Munk School," he said.) They argued that without a Canadian victim of real consequence not much could be done, and that the mechanisms put in place by Koobface to generate revenue were so subtle that it was extremely difficult to identify who the victims were. Although Koobface netted millions a year, the earnings were derived from hundreds of thousands of micro-transactions, a fraction of a penny each, spread across dozens of countries. Furthermore, without an identifiable complainant it is almost impossible for a police force to justify the resources to investigate a case like Koobface. Police officers ask, "What's the crime?" Prosecutors ask, "Who am I supposed to prosecute?" Koobface, it appeared, would fall through the cracks.

Cyber-crime networks, especially international ones, succeed by hiding locally while leveraging the global infrastructure of a free and open Internet. Electrons may move at the speed of light, but legal systems crawl at the speed of bureaucratic institutions, particularly across international borders. We told the RCMP that several of the major command-and-control computers used by Koobface were rented out on servers in Britain and Sweden, and that the perpetrators might be out of reach in St. Petersburg but these surely could be seized. "For us to get permission just to talk to a counterpart in the United Kingdom or Sweden could take months," we were told, the sense of resignation obvious.

The RCMP officers told us they would explore the case further, but left us with the distinct impression that what they would actually do (or not do) was none of our business. "We'll take it from here," was all they said. While they did not ask us to withhold publication, knowing that doing so would be inappropriate, they did suggest that our report might prejudice their investigation. We told them that we had an obligation to publish and gave them a realistic time frame in which we would do so. Our report, *Koobface: Inside a Crimeware Network,* went live in the fall of 2010.

The outreach with the RCMP was one track we followed, but we also worked with the broader security community to notify the hosting companies and ISPs that serviced the roughly 500,000 fraudulent Google Blogger and Gmail accounts and the tens of thousands of Facebook pages upon which Koobface had built its malignant enterprise. Doing so gave us a window onto a different kind of cyber-crime enforcement performed by private sector companies taking matters into their own hands. Many were increasingly frustrated with the slow pace and awkward political constraints around official cyber-crime responses and had begun to find ways to dismantle or degrade criminal networks and botnets on their own. Specialists working for Facebook, Jan Droemer and other security researchers (notably Dirk Kollberg of the company SophosLabs and independent security consultant Dancho Danchev) continued their pursuit of Koobface for more than a year, culminating in a dramatic January 2012 outing of the perpetrators first by Danchev, then Facebook, and finally Droemer and Kollberg in a detailed report published by SophosLabs that revealed reams of personally identifiable information about the group. The public exposure and the release of the Sophos report led to immediate action by Koobface: its command-and-control servers stopped responding, and the gang started removing traces of themselves from the Net. The antivirus company F-Secure called it a "name and shame approach" – one that was widely criticized by some in the industry for hampering an ongoing criminal investigation and jeopardizing the collection of evidence.

With their identities revealed, and their infrastructure brought to its knees, Koobface will not be able to operate with the same carefree impunity it once did, but it is unlikely its creators will ever be prosecuted. Russia lacks extradition treaties with the U.S. and other Western countries, and the arrest and prosecution of the group is not likely there. Recent history suggests that Russian cyber criminals have little to fear as long as they stay close to home. (Responding to the Koobface incident, Russia's anticyber-crime unit, the interior ministry's K Directorate, told Reuters that it did not investigate the matter because it had not been asked to: "An official request needs to be filed to the K Directorate first, and when it's filed, we will certainly investigate and work on it." Officials at Facebook told the same Reuters reporters that they had passed along information to the interior ministry before deciding on their more radical naming and shaming approach.) In February 2011, in another case, a Russian criminal, Yevgeny Anikin, received only a suspended sentence after being arrested for what American authorities called "perhaps the most sophisticated and organized computer fraud attack ever conducted," a hack of the Royal Bank of Scotland and a $9 million windfall for Anikin.

• • •

Ever since the Internet emerged from the world of academia into the world of the rest of us, its growth trajectory has been shadowed by a grey economy that thrives on opportunities for enrichment made possible by an open, globally connected infrastructure. In the early years, cyber crime was clumsy, consisting largely of extortion rackets that conducted network attacks against online casinos or pornography sites to extract funds from frustrated owners. Koobface is part of what author Misha Glenny calls the "industrialization of crime on the web."

In the early days, cyber crime was primarily a loner's calling, an annoying but affordable by-product of an open Internet. Today, the loners find each other, network together, and professionalize their activities. Underground forums have emerged in the dark recesses of the Internet where specialized tools and techniques are now bought, sold, and traded. Malicious software packages – known as "0days" or "zero days," because antivirus companies have no known protections against them – are now as readily available as songs on iTunes. "Botnet herders" – individuals who control tens of thousands of compromised computers – market their wares in underground auctions. Stolen credit cards and email addresses are sold, bought, and traded like candy. (Rik Ferguson, of the Internet security firm Trend Micro, provides a detailed list of illicit products and services sold. To name a few: hiring a DDOS attack, $30–$70 a day; hacking a Facebook or Twitter account, $130; hacking a Gmail account, $162; scans of legitimate passports, $5 each.) Around the globe, botnets can be rented cheap online from public websites for weeks, days, even hours. Some advertise 24/7 technical support. Cyber crime has indeed become a global menace, a multinational business that shows no signs of letting up, a former cottage industry gone viral and into a global marketplace.

Whereas ten years ago a cyber criminal needed the equivalent of an advanced graduate degree in engineering, today a teenager could set up something like Koobface. In Brazil, there is an academy that openly advertises courses on computer crime: "This course is intended for everybody making online transactions. You will learn how crackers take control of corporate or home computers . . . how 'auto-infect' works, how to use sources [trojans], how to manipulate the security plug-ins installed on browsers such as Internet Explorer, Firefox, Chrome, Avant, Opera, and antivirus and firewalls. How spamming helps catch new victims, what 'loaders' do and how crackers use them . . . how crackers can own e-commerce websites that store credit card numbers and what they do with this data. You'll learn about the laws in Brazil and what the sentence is if you're caught."

The course costs $75 and includes a special bonus: 60 million email addresses with which to begin experimenting. (Brazenly, the academy lists its office address, and phone and fax numbers on a public website with an accompanying Google map location.) But then again one needn't go to cyber-crime school, or pay any kind of fee at all. One freely downloadable program provides a simple click-as-you-go interface to create "phishing" websites that simulate legitimate banking, shopping, and webmail interfaces, but which are actually designed to extract credit card numbers, email addresses, and passwords from unsuspecting victims. Just follow the step-by-step screen instructions guiding you through how to create a mock site, load it online, and then send links out to potential victims.

Cyber crime thrives not just by its ingenuity, but also by social media opportunities. Koobface succeeded by mimicking normal social networking behaviour. It leveraged our readiness to extend trust with our eagerness to click on links in a world that has become intensely interactive. The age of mass Internet access is less than twenty years old, and social networking, cloud computing, and mobile connectivity are, for most people, innovations only of the last few years. We have embraced these new technologies at such a pace that regulatory agencies have been left in the dust, and we have overlooked extraordinary user vulnerabilities. Today, data is transferred from laptops to USB sticks, and over wireless networks at cafés, and stored across cloud computing systems whose servers are located in far-off jurisdictions. We produce massive amounts of personal data as we navigate this new ecosystem and click on website addresses and documents like lab mice clicking on pellet dispensers. It is this conditioned tendency, combined with the sheer volume of data we generate, that Koobface and others capitalize on with precision. Every new piece of software, social networking site, cloud computing system, or web-hosting service represents an opportunity for the predatory cyber criminal to subvert and exploit.

• • •

Cyber crime has become one of the world's largest growth businesses. (Estimates vary, and the self-interest of threat inflation cannot be ignored, but the National Security Agency's General Keith Alexander has estimated that American companies lose around $250 billion from IP theft, and that internationally cyber crime causes $114 billion in losses. The computer security company McAfee states the number is closer to $1 trillion. Whatever it is, one thing is clear: it's large.) This growth is being fuelled by the demographic changes affecting the entire Internet. Russian, Chinese, and Israeli gangs are now joined by upstarts from Brazil, Thailand, and Nigeria. Executing a digital break-in of a computer in Manhattan can be done from the slums of Panama City.

Western observers of technology tend to have a biased view of what constitutes digital innovation. We think of ingenuity in ways that conform to our concepts of what is right and moral: liberation, freedom, commercial entrepreneurialism for the common good. We think about our kids creating avatars for their online games, or a new iPhone app that shows us the location of the nearest laundromat. But ingenuity in cyberspace is as bountiful and unpredictable as the individuals who go online, and for the newly connected digital natives of the global South and East operating an email scam or writing code for botnets, viruses, and malware represents an opportunity for economic advancement, a relatively safe route around structural economic inequality and mass unemployment, an avenue for tapping into global supply chains and breaking out of local poverty and political inequality. Sitting in front of a glowing monitor thousands of miles from their victims, essentially immune from the law in St. Petersburg or Rio de Janeiro, scamming online must feel more like a virtual crime, with very tangible monetary rewards.

The speed by which new cyberspace technologies are created far outstrips the capacity of governments to regulate them. As William Gibson argues, "In a world characterized by technologically driven change, we necessarily legislate after the fact, perpetually scrambling to catch up." Our law enforcement is structured around a world of sovereign states, each separated from the other in territorially based pyramids of hierarchical authority. Cyber crime, on the other hand, is global, its victims distributed across political jurisdictions, and it cleverly avoids concentrations high enough in any one jurisdiction to warrant concerted law enforcement attention. As our Koobface experience showed, even armed with a blueprint of the group's entire infrastructure, the RCMP faced a mountain of bureaucratic hurdles working with counterparts abroad.

• • •

An epidemiologist studying such intense sharing and interaction introduced into a population with no prior immunity would predict the digital equivalent of a virulent disease outbreak. And that's just what has happened. How big is cyber crime? Statistics are regularly thrown out by governments and security companies, but they are inherently unreliable. Still, as the *Economist* recently put it, "Big numbers and online crime go together." The security firm McAfee estimates that they receive 80,000 new malicious software samples a day; that is, 80,000 new forms of malware that they have never registered before, newly created, every single day. But the reality is that no one really knows the full extent of the maliciousness. As Misha Glenny, author of *DarkMarket*, puts it, figures like these could be "wildly exaggerated . . . by the cyber security industry in order to generate sales. Or it could be the result of some hyperactive algorithms. Or it could be true. But nobody can assert with any confidence which it is." New major hacking attacks, data breaches, and other forms of cyber crime come to light almost daily.

Consider the last week of December 2011. As celebrants were revelling in the holiday season and looking forward to ushering in the New Year, two dramatic back-to-back breaches occurred in a span of two days. First, a December 27, 2011, breach of the Chinese blogging site Tianya resulted in the exposure of 46 million users' email addresses and passwords, with 4 million of the usernames and passwords published entirely in the clear; that is, not encrypted. Two days later, the hacking collective Anonymous breached the website of the private security think tank Stratfor and dumped 860,000 accounts, including usernames, email addresses, and passwords, into the public domain. The hackers claimed to have used the credentials of thousands of credit cards collected during the hack to make fraudulent Christmas donations to human rights groups and other NGOs.

The Christmas donations made by Anonymous underscore another characteristic of cyber crime, and a final thought about the Koobface story that should give us pause. The ingenuity of online crime is often combined with humour and irony. After the security company Symantec issued a 2011 report that included details on a particularly malignant backdoor trojan horse called Poison Ivy, used by attackers who had infiltrated more than fifty companies, the perpetrators, a group called Nitro, began sending out emails purporting to be from Symantec with the Symantec report attached. The report was infected with the Poison Ivy trojan itself! One hacker goes by the name "Google," making it almost impossible to locate him or her using open Internet searching methods. (Try searching for "Google" on Google.)

The Koobface group was notoriously humorous, often charming, and even had what might be considered a kind of ethical restraint: they openly mocked security researchers who tracked them, but in doing so also inadvertently exposed the limits of their desire *not* to do major harm. For example, in 2009 Koobface left a Christmas greeting for security researchers that attempted to clarify their intentions: "As many people know, 'virus' is something awful, which crashes computers, steals credential information as good as [sic] all passwords and credit cards. Our software did not ever steal credit card or online bank information, passwords or any other confidential data. And WILL NOT EVER." In other words, they restricted their crimes to petty fraud, albeit on a massive scale. The alarming thing is that they could have easily done otherwise.

Hundreds of thousands of compromised computers networked together through invisible strings controlled by a few individuals can be employed to extract pennies from unsuspecting victims, as it was with Koobface, or sensitive national security documents from government agencies, as GhostNet and Shadows proved. Such a system can be used to direct users to click on fake advertisements for Viagra, or be marshalled to attack a human rights website, as happens with increasing frequency from Iran and Kazakhstan to Burma and Vietnam.

Lurking in the background is another disturbing question: What happens when the world of cyber crime becomes militarized?

9.

Digitally Armed and Dangerous

"**Listen,**" said the Indian official, the connection echoing, crackling with static, "we would like to know whether you would join us."

"Excuse me?"

"We want you to help us fight back. Is that something you would be prepared to do?"

"What do you mean, 'fight back'?"

"Yes, yes, fight back. Join us to attack the Chinese."

We looked at each other in stunned silence. Were we actually hearing this?

The Skype call had been hastily set up between the Citizen Lab and an obscure Indian state intelligence organ called NTRO in the spring of 2010. An Indian version of the U.S. National Security Agency, the National Technical Research Organisation might sound humdrum, but it is India's highly secretive premier technical intelligence-gathering organization. NTRO stands at the apex of the Indian armed forces and secret services and, among other things, is responsible for satellite, electronic, and Internet monitoring activities. Judging by the offer extended to us over Skype, NTRO is not above asking for outside help, even outsourcing work necessary to getting to the root of attacks on the computer networks it polices, especially when it believes that counterattacks are required.

While dozens of embassies, prime ministers' offices, and diplomatic missions around the world had been plundered by the Ghostnet attackers, the perpetrators described in *Shadows in the Cloud* instead focused like a laser beam on the Indian national security establishment. As in the past, mistakes made by the attackers gave us insight into their inner workings, but in this case we were able to recover copies of the data being removed, or "exfiltrated" as they say in the cyber intelligence world.

The Shadows attackers went to great lengths to obscure their trail, splitting the documents and other data stolen from unwitting computer owners into bits and pieces and hopscotching them across the Internet through "drop zones" set up on a spider's web of free hosting sites, before reassembling them. Nart Villeneuve was able (again) to get partial access to one of these stepping stones – an open file transfer protocol (FTP) used by the attackers on an improperly secured computer – and once he found this window into their subterranean lair, he engineered a script that automatically copied anything that passed through the FTP site. As I told John Markoff of the *New York Times* (which gave front-page coverage to *Shadows in the Cloud*), we were going "behind the backs of the attackers and picking their pockets."

As with the GhostNet investigation, we had privileged access to Tibetan computers, including those situated in the Office of His Holiness the Dalai Lama in India, which we had wiretapped with permission. Seeing it from both sides – from that of the victims (the Dalai Lama's office and the Tibetan Government-in-Exile), and that of the attackers (through a backdoor left open into their networks) – allowed us to confirm that the data being exfiltrated was, in fact, stolen from the source. Part of the data was a folder called "Letters," which contained a year's supply of the Dalai Lama's official correspondence. Replies and inquiries to numerous individuals, organizations, and world leaders, and to their governments all silently stolen from the Dalai Lama's office computers were now in our possession *and* that of the unknown attackers.

We also gained access to hundreds of other documents processed through the drop zone – spreadsheets, PowerPoint presentations, hotel reservations, expense reimbursement forms, and other letters and charts – that seemed unrelated to Tibet or the Dalai Lama. We reassembled what could be recovered, assigned a code to each piece of data, and tried to determine who, other than the Dalai Lama, the attackers were pilfering. Some of the documents appeared highly sensitive: two were stamped "Secret," six "Restricted," five "Confidential," and one appeared to be diplomatic correspondence. By combining forensic analysis of the documents' metadata with IP addresses and other open-source information, we determined that the sensitive information was stolen from at least one member of the Indian National Security Council Secretariat, probably the Indian Directorate of Military Intelligence, and others in Indian government consulates and embassies, including the Indian embassy in Washington.

Some of the documents were extraordinary: secret assessments of India's security situation in flammable states like Assam, Manipur, Nagaland, and Tripura; security reports about insurgency groups threatening India, such as the Naxalites and Maoists; and confidential information taken from Indian embassies about the country's relations with and strategy towards West Africa, Russia and the rest of the former Soviet Union, and the Middle East. We pieced together visa applications, passport office internal memos, and diplomatic correspondence. Among the stolen documents were visa applications made by Canadians at the Indian embassy in Kabul. Few Canadian tourists travel through Afghanistan en route to India these days, but in 2010 Canada was actively participating in the NATO mission there. Were the visa applications from Canadian military personnel?

Defence-oriented Indian academics and journals had also been compromised and, while none of the material obtained was classified, the documents we recovered pertaining to these individuals and organizations revealed information about very sensitive topics. (This suggested that the attackers had managed to compromise individuals knowledgeable about classified information.) We recovered documents relating to missile programs and artillery combat command-and-control systems. Ironically, some of the documents contained references to "network centric warfare," including cyber espionage threats (and how to defend against them) made against the owners of the reports.

The highest intelligence and diplomatic organs of one of the world's largest governments had been thoroughly exposed and penetrated by hackers, and only two groups of people knew about it: us and them – and the other guys, we presumed, were operating at the behest of the Chinese government.

• • •

In the days before cyber espionage, acquiring this amount of detailed intelligence would have required risky, time-consuming, and laborious missions involving physically infiltrating buildings and compromising personnel with clearances and access to data. Even then, it is hard to imagine a single operation siphoning up (in one elegant electronic scoop) such a wide array of information. More remarkable still was that the cyber espionage was continuously occurring right under the noses of the victims. The perpetrators were able to receive minutes of Indian National Security Council meetings as they were transcribed into Word documents and saved on the compromised computer. In short, we witnessed an international spying operation as it unfolded, in real time.

Seeing first-hand evidence of what had been stolen added a

unique dimension to this investigation that was missing from the GhostNet probe. The latter was larger in scope, and perhaps more debilitating to the victims because of the Ghost RAT trojan – which allowed the hackers to remotely turn on the audio and video capture systems of computers under their control and use them to silently eavesdrop – but we never saw what the GhostNet attackers actually stole from victims. Our conclusions were mostly based on inference. With Shadows we had tangible evidence of what was being removed.

As exciting (and unprecedented from a research perspective) as this was, it left us in an ethical quagmire. What should we do about possessing data of a foreign government that is marked classified, sensitive, and restricted? There is no training for this type of situation in academia, no textbook on the handling of a foreign government's classified material silently recovered by a clandestine cyber espionage ring. As with the GhostNet and TOM-Skype reports, we found ourselves in *terra incognita*.

After debating several options, we felt obliged to present a detailed brief to the Indian government so that it could patch the gaping holes in its computer networks and prevent further exploitation by other cyber spies. The call was set up by Steven Adair of the Shadowserver Foundation, one of our partners in the investigation. Shadowserver consists of a group of U.S.-based volunteer computer security professionals, many with extensive contacts in the law enforcement and intelligence communities, and it was able to make contact with an official at NTRO. We sent the official a spreadsheet listing the IP addresses of the computers we knew were compromised and an overview of the documents we possessed, and asked him for guidance on how to dispose of them. Instead, to our astonishment he offered us a job: help NTRO "attack the Chinese."

While the request to hack into Chinese government computer networks startled us, we weren't surprised by the Indian government's instinct to "hack back," or its willingness to outsource the job. Alongside other governments, India was groping with how to defend itself from persistent cyber attacks, and, like many of them, it followed the basic adage about "offence being the best defence." (Indeed, by June 2012 the *Times of India* was reporting from unnamed sources that "the National Security Council (NSC) headed by Prime Minister Manmohan Singh would soon approve [a] comprehensive plan and designate the Defence Intelligence Agency and National Technical Research Organisation as agencies for carrying out offensive cyber operations, if necessary.") But unlike the United States, Britain, and other Western democracies, the Indians, we sensed, were not going to follow a conventional playbook should they decide to fight back.

Had we pursued the offer to "fight the Chinese" we would have no doubt enjoyed a lucrative and exciting, but highly unethical, set of missions. It was not the first time – and it would not be the last – that the Citizen Lab had to decide whether or not to intervene in the very world we were studying, and nor was it the first time we were confronted with the blurring worlds of cyber crime, espionage, and international warfare.

• • •

Another day, another hacker exploit. Only this time the perpetrator was not Anonymous or LulzSec or any of their hacker sympathizers. In February 2012, a group called the Syrian Electronic Army (SEA) posted on Internet forums the email credentials, including usernames and passwords, of Al Jazeera journalists, as well as a series of emails purporting to show bias in their coverage of the Assad regime. We learned about this breach the same way most other concerned observers did: when the SEA boasted about it on their Arabic Facebook page. That's right, the SEA has a

Facebook page. In fact, the pro-regime cyber warriors have set up hundreds of them, and as soon as Facebook administrators receive a complaint about the group violating its terms of service – for inciting violence or using Facebook to disseminate links to malicious software – the offending page is removed, prompting the SEA to simply create a new page with a new domain name. (At the time of writing, the latest domain was http://www.facebook.com/SEA.P.187.) It is an online version of Whac-A-Mole only in this case it's not a game, it's war.

The SEA also has a Twitter account, through which posts are made in Arabic that taunt its adversaries or boast about its latest exploit. For example, on July 5, 2012, the SEA managed to take over the Twitter account of Al Jazeera's *The Stream* – possibly acquiring the sign-up credentials through a previous computer breach of Al Jazeera's servers – and then took credit for the hack on its Twitter account, @Official_SEA. For a few hours on that July day, to the bemusement of many Twitterati, they used Al Jazeera's account to turn the broadcaster's coverage upside down: from an independent monitor of atrocities to a mouthpiece for the Assad regime.

The Citizen Lab turned its attention to the SEA when the Arab Spring blew into the streets of Damascus in early 2011. Amidst the smoke and rubble of an increasingly violent civil war – and after the UN monitors finally reported that "crimes against humanity" were being committed by the Syrian regime – another type of warfare took shape, this one through radio waves and fibre-optic cables, and over social media platforms.

Like the Tunisians, Egyptians, and Libyans, angry Syrians opposed to the dictatorial ways of their government and looking to ignite a revolution reached instinctively for the latest tools of the digital age. The anti-Assad "Day of Rage," announced to the world through Arabic Facebook, Twitter, and on other social media platforms in February 2011, set the tone. The Syrian protesters built on lessons learned from other digitally empowered protests, and benefited from a growing grassroots movement of technological peer support. Hacktivist groups like Telecomix and Anonymous jumped into the fray by breaking into Syrian government computers, distributing secure tools to circumvent Internet censorship, and helping expose companies that provide services to the Assad regime. In February 2012, Anonymous broke into the email server of the Syrian Ministry of Presidential Affairs and published hundreds of emails. As usual in such domestic conflicts, neighbouring states and great powers meddled in this one, too. While Russia and China stymied UN resolutions to sanction Syria, Iran's Revolutionary Guard's elite signals intelligence unit roamed Syrian city streets in black vans and employed sophisticated surveillance tools to triangulate the location of dissidents using insecure satellite phones. On the other side of the battle, American and British officials provided tools and training for the armed opposition in the Free Syrian Army, while the Canadian government quietly used its diplomatic headquarters in Ankara, Turkey, to channel information to those fighting the Assad regime.

As a result of such outside support, those opposed to Assad are technologically well equipped. The latest generation mobile phones have been employed as frontline sensors, uploading atrocities for the world to witness as they occur – their shaky, hand-held videos a grim portal into the otherwise hidden spectacle of torture, suffering, and death – thus circumventing the Syrian regime's official blackout of journalists. The Citizen Lab's senior Middle East and North Africa–based researcher, Helmi Noman, has shared many of these these videos with our Toronto staff, translating the horrific scenes from Arabic to English so that we could understand that protesters were being buried alive at gunpoint, forced to swear allegiance to Assad while they drew their last breath; that tidy lines of corpses covered in blood-stained white sheets, some clearly children, were the victims of deliberate Syrian military attacks on the country's own people in its own cities.

But the familiar script of digitally enabled pro-democracy activists outflanking flat-footed tyrants, which played itself out in other theatres of the Arab Spring, never fully materialized in Syria. The Assad regime adapted and evolved, taking its counter-insurgency tactics to the virtual plane. Unlike the leaderships of Egypt and Libya, who in last-ditch acts of desperation pulled the plug on the Internet, after various ham-fisted attempts at control, Syria decided instead to actually *loosen* its grip on cyberspace. Facebook, Blogspot, YouTube, and Twitter, perennially censored by the xenophobic regime, were suddenly made available at the very moment activists took to the streets and to their mobile phones. A conciliatory gesture perhaps? An appeasement to the protesters' demands for more free speech and access to information? More likely the powers-that-be had a more sinister strategy in mind.

Part of that sinister strategy involves surveillance. By loosening controls over particular Internet platforms – especially those used by protesters to organize – the Syrian regime acquired unparalleled insights into its adversaries' thoughts, plans, and actions. As the conflict unfolded, reports began to surface about a dark market in high-tech equipment – the products and services coming mostly from Western firms – used by the regime. In a series of investigative reports, Bloomberg News revealed that an Italian company, Area SpA, was installing a surveillance system that would enable the Assad regime to intercept, scan, and catalogue emails flowing through the country. The report was the tip of an iceberg.

The Citizen Lab helped uncover that routers belonging to Blue Coat Systems, an American company based in Sunnyvale, California, were widely deployed across the Internet in Syria. Our researcher Jakub Dalek discovered the Blue Coat devices by running a series of specially designed network scans, the equivalent of a digital flashlight searching through the sewers and catacombs of Syrian Internet space looking for fingerprints of specific equipment used. The Blue Coat devices could be used to filter content and monitor communications in fine-grained detail. Under U.S. sanctions against the sale of products and services to Syria – designated a "state sponsor of terror" by the American government – any business relationship between Blue Coat and Syria was illegal.

The European hacker collective Telecomix was on the same trail as the Citizen Lab, and published reams of unfiltered data they had collected about Blue Coat. Our report was released a few days later, on November 9, 2011, and both reports led to a firestorm, including calls for a U.S. Congressional investigation into Blue Coat. The company later acknowledged the presence of their devices in Syria, but said they were shipped to the country fraudulently and without their knowledge, a dubious claim. As Blue Coat's primary function is to monitor Internet traffic, and their devices only function properly when checking in to get updates from central Blue Coat servers, such a claim was too far-fetched to be credible. These and other revelations of high-tech surveillance equipment being imported into Syria underscored the other side of a regime that once attempted to control the Internet through censorship: targeted surveillance is far more effective.

Just as the Citizen Lab was preparing its Blue Coat report, we stumbled upon a number of Syrian government websites that were hosted on Canadian servers, including the state-backed television station, Addounia TV, that had been placed on an official sanctions list by Canada and the European Union for incitement of violence. The content being streamed online by Addounia TV claimed that the atrocities captured on film by Syrian protesters were fabrications, and it encouraged Syrians who supported Assad to take to the streets and fight back. In a bizarre twist Addounia was hosted on computers located in Montreal, and we also found that the website of Al-Manar, the media wing of the Lebanese militant group Hezbollah, was hosted on the same Montreal-based servers, again in violation of Canadian sanctions. Reflecting on the role media have played in inciting genocide in places like Rwanda, we decided to publish our findings immediately. Called *The Canadian Connection: An Investigation of Syrian Government and Hezbullah Web Hosting in Canada,* our report no doubt caused a few red faces in Foreign Affairs and International Trade Canada, but it also underscored the complexity and difficulty of imposing effective international sanctions over cyberspace activities. Nonetheless, believing that web hosting constituted "material support" for the Syrian regime and Hezbollah, we chose to act swiftly.

● ● ●

High-tech surveillance equipment in Syria and Syrian government web hosting in Canada were only part of the story of Syria's metamorphosis from an Internet-phobic regime to one that embraces technology in the service of armed struggle and civil repression. The SEA's first forays into cyber war may have been amateurish – it defaced websites, the online equivalent of graffiti; spammed the comments sections of online forums and newspapers, the actions of a pest more than a menacing army; and targeted websites and forums that appeared to have no relation whatsoever to Syria (the website of an obscure town council in Britain, Harvard University, and so forth), juvenile acts of opportunism. Website defacements of this sort demonstrate a low level of expertise: anyone with a few hours to spare can easily Google instructions and then scan the Internet looking for low-hanging fruit, poorly patched servers waiting to be plucked and desecrated. But over time, and especially into 2012, SEA evolved, its methods becoming increasingly sophisticated.

In the spring of 2012, the Electronic Frontier Foundation started receiving reports from inside Syria of phishing attacks on Facebook, YouTube, and other social media outlets used by Syrian dissidents. The EFF found that when users clicked on links posted on the comment sections of opposition Facebook and YouTube sites, they were taken to fake websites that encouraged them to download special software, which was then used to acquire their credentials and sometimes to take over their computers. The EFF also discovered an instance of a malicious software program hidden in images circulated among Syrians in the diaspora. Although EFF could not confirm the identity of the perpetrators, they suspected that the Syrian telecommunications ministry was behind the attacks. Meanwhile, reports of authorities using force against activists and dissident Facebook users, and demanding their login information, surfaced. In one case, a user was beaten by Syrian police, who then informed him that they had been reading his "bad comments" on Facebook. After providing his password to authorities, he was imprisoned for two weeks. Upon his release, he found that somebody had logged into his Facebook account and posted pro-regime comments in his name.

Google computer security analyst Morgan Marquis-Boire and UCLA Ph.D. student John Scott-Railton were involved in the EFF's work, and in 2012 they contacted the Citizen Lab to suggest combining research efforts with EFF's Eva Galperin. (Marquis-Boire and Scott-Railton later joined the Citizen Lab as research fellows.) Together, our teams have uncovered one targeted attack after another on Syrian dissidents, typically engineered by commandeering someone's computer and using that person's Skype or email account to trick the dissident's network of contacts into clicking on links or opening files that contained malicious trojans. Whereas prior defacement and spam attacks had the imprecision of a sledgehammer, these attacks were more like carefully calibrated pliers. Our researchers watched as the cyber raids became more persistent and sophisticated, using several commercial remote administration tools bundled and hidden in malicious software, which suggested significant knowledge of criminal hacking techniques. When the author of one of these tools, called Dark Comet, discovered through our published reports that his software was being repurposed by the SEA to trap dissidents, he was horrified, issued an apology, and announced that he would no longer maintain the software as a freely available product. This did little to slow down the SEA. Within days there were more attacks targeting Syrian dissidents, this time using a different commercial remote administration tool called Blackshades.

Although we found no smoking gun connecting these attacks directly to the Syrian government, the majority were clearly engineered by individuals connected to command-and-control computers operating on Syrian telecommunications networks registered in Damascus. A Citizen Lab contact with extensive dealings in the domain registration business gave us a likely set of names and Syrian-based cellphone numbers connected to the names and email addresses used to register the domains linked to the attacks, but we decided not to publish them for fear of endangering lives. Clearly, though, the Syrian government was either tacitly condoning or actively encouraging the SEA, a marked turning point in how an autocratic regime deals with a digitally mobilized opposition. Dictators have little to fear from technology: it can be their best friend.

Syria's SEA is a curious hybrid. Not formally linked to the Syrian government, it nonetheless undertakes information operations in support of the regime, and does so at arm's length so as to ensure plausible deniability. Its methods are not technically complex; indeed, they are run-of-the-mill and widely employed in the world of cyber crime, and they are attractive because they are cheap, easy to use, and often enough extremely effective. This is precisely what makes the SEA case noteworthy: the methods, tools, and tradecraft of cyber crime are being repurposed and deployed by one of the world's most repressive states in the midst of a bloody civil war, a new model of "active defence" emerging among autocratic regimes the world over. The exploitation of cyber-crime techniques is an increasingly common state-sponsored form of military action in cyberspace, and the already percolating menace of cyber crime is morphing into a boiling cauldron of espionage, sabotage, warfare, and repression.

· · ·

Among those governments using cyber-crime techniques for national military and intelligence purposes Syria may be the most recent, but it is not the first nor the most voracious. That title goes to China, whose adversaries have been the most frequently targeted, and for the longest periods of time. China has used just about all of the latest techniques of the cyber-criminal underworld for strategic intelligence, industrial espionage, and military action. Indeed, it is fair to say that China is *the* template for state-sponsored cyber crime.

During Ghostnet, basic Internet "social engineering" techniques – the art of fooling people into divulging confidential information – first refined by cyber criminals were used to fool recipients of emails at the Office of the Dalai Lama and Tibetan Government in Exile into opening attachments that contained a very simple piece of malicious software. Once infected, the attackers installed a more sophisticated remote administration tool on their computers, a freely available and open-source piece of software known as Ghost RAT (hence the name of the espionage network). During Shadows, the attackers borrowed from the widely deployed criminal method of splitting up and routing stolen documents from victims' computers across redundant social networking platforms to ensure resiliency and to disguise the origins of the malicious network in case parts of their infrastructure were reported on and shut down. When members of the Foreign Correspondents' Club of China were targeted by socially engineered emails containing malicious trojans, the infected computers connected back to Taiwan-based command-and-control servers under the control of the attackers. (The compromised servers were based at Taiwan University and were the very ones used to distribute antivirus software to staff and faculty.) When the European Parliament passed a resolution condemning China's repression of Tibet, the text was immediately repurposed to contain a malicious piece of software and then distributed to the contact list of an exiled Tibetan whose computer was compromised by Chinese attackers. When Twitter was used as a means to raise awareness by Tibetans about an important anniversary, pro-regime hackers employed several hundred bots – automated programs that generate content – to flood Twitter discussions using the hashtags #Tibet and #Freetibet, making those hashtags unusable, a technique known as "hashtag bot-flooding" originally developed by spammers. Chinese hackers redeployed a common technique, an iFrame injection, or "drive-by" attack, in which the websites of their adversaries are hacked into and loaded with malware that targets visitors using improperly secured browsers. Over the years numerous websites of prominent human rights groups have been exposed in this manner, including Amnesty International U.K. and Human Rights in China.

In each case, many of the primary methods and tools used were not specially designed or custom built; instead, they were simply repurposed from the world of cyber crime, and many observers believe China tacitly condones and supports the vast cyber criminal underworld because it benefits from it. Looking at the evidence, it's hard to conclude otherwise. And China is not alone. Many shadowy underground entities employ cyber-criminal methods against human rights and opposition groups in operations that benefit entrenched authorities. Russia, Kyrgyzstan, Belarus, and other states across the former Soviet Union represent good examples.

In February 2005, during parliamentary elections in Kyrgyzstan, websites belonging to political parties and independent media aligned with the opposition were subjected to unexplained technical failures, glitches, and deliberate hacking. Journalists at independent media organizations had their email accounts flooded with large volumes of spam and phony emails. Several websites were hacked and defaced, and one had its domain name deregistered because the authorities claimed it had no "legal status." Shortly thereafter, a major DDOS attack, undertaken by a group calling itself Shadow Team, overwhelmed Kyrgyzstan's leading ISPs. OpenNet Initiative's Krygyzstan-based researchers obtained the extortion note sent by Shadow Team to the ISPs, which threatened to continue the attacks until specific websites connected to the political opposition were shut down. A separate threatening email was sent to a popular regional news site, http://www.centralasia.ru, demanding that it stop publishing any and all information about the situation in Kyrgyzstan. The perpetrator turned out to be a single hacker operating out of Ukraine, but whose attacking computers were physically located in the United States. The same hacker was simultaneously pursued for different reasons by U.S. security researchers, and eventually the botnet was disabled.

Based on ONI's experiences in Kyrgyzstan, leading up to the 2006 Belarus presidential elections we assembled a group of researchers (both inside and outside the country) to monitor the Internet. Although ONI testing indicated that Belarus, like Kyrgyzstan, had no Internet censorship, the regime of President Alexander Lukashenko was (and still is) widely considered typical of Soviet-style authoritarianism: prone to silencing dissent and quelling opposition using heavy-handed methods. Indeed only a year before, Ilya Mafter, the program officer from the Open Society Institute (which had funded the ONI project) had been arrested on trumped-up money laundering charges and held in detention for several months. Before and during the presidential election, ONI documented numerous opposition websites coming under denial-of-service attacks, or made inaccessible on the state-owned Beltelecom network. During a day of major demonstrations in the capital city Minsk, when riot police intervened to disperse and arrest protesters, one of the main dial-up services for Internet connectivity in the city went dead, having experienced "technical problems."

In 2008, two years after the Belarus election, war broke out between Russia and Georgia over the disputed territorial enclave of South Ossetia. As Russian tanks stormed the territory, ONI researchers inside Georgia and neighbouring countries monitored the information domain, collecting evidence of computer network attacks and filtering. At the war's height, Georgian government websites and much of its information infrastructure, including banking and emergency services, came under a massive denial-of-service attack, which most people attributed to the Russian government. (A similar assault had been inflicted on Estonia a year earlier, when that country's leaders made the unpopular decision to relocate the *Bronze Soldier of Tallinn*, an elaborate Soviet-era war memorial, along with the remains of Soviet soldiers.) Desperate to stem the attacks, and hoping to counter Russia's disinformation campaign, the Georgian government censored access to all Russia-based websites. Accustomed to seeing Russian news online, and unaware of the decision taken by their government, Georgians in the capital city of Tbilisi panicked, fearing that the blackout pre-saged a massive Russian ground assault. Rumours quickly spread of tanks approaching the outskirts of the city.

Were the DDOS attacks orchestrated by the Russian military, undertaken by sympathizers to the Russian cause, or some combination of the two? No one would or could tell. To illustrate how easy it is for anyone to participate in such attacks, journalist Evgeny Morozov, writing for *Slate* magazine, downloaded instructions for one of the DDOS tools advertised on Russian forums and, in less than an hour, was a participant in the attacks on Georgian government websites himself.

After the war ground to a halt, Citizen Lab researchers were able to register the domains of the botnets responsible for the DDOS attacks, which the owners had let expire. Doing so gave us a precise sense of the breadth of commandeered computers under the hackers' control during the Russian–Georgian conflict, as the zombie computers still "checked in" with domains now under our supervision. While most observers talked about Russian-based attacks on Georgian government websites, we found instead a global network of zombie computers used to assault the Georgian infrastructure, the vast majority of which were physically located in the U.S. and Germany. We also determined that the same botnets had been used in numerous recent criminal activities, mostly involving extortion against pornography and gambling websites.

The worldwide distribution of computers linked together to assault Georgia proves how difficult it is to "attack back" those causing mayhem in cyberspace. Indeed, at one moment during the conflict, when the Georgians took up an offer from a Georgian ex-pat based in Atlanta to host their websites in the United States, commandeered U.S.-based computers were overwhelming other U.S.-based computers hosting Georgian government websites!

• • •

What ONI researchers have found in the former Soviet Union has parallels in other parts of the world. In 2009, the Citizen Lab analyzed DDOS and defacement attacks that were vexing the Burmese opposition and independent media outlets alike. Most observers, including the victimized organizations themselves, blamed the Burmese government, but Nart Villeneuve determined that the attackers had no ties at all to the Burmese government. Instead, the attacks had been launched by a group of Burmese hackers trained as computer programmers at Russian military academies in Rangoon. Overseas Burmese pro-democracy groups had apparently irritated them, and they took it upon themselves to defend the military junta by menacing the groups persistently over the Internet. Part of the hackers' motivation was to earn bragging rights, and their undoing was that they boasted about their exploits on chat forums that we were monitoring, allowing us to triangulate their usernames with other coincidental pieces of information. Still, why shut down cyber crime in your backyard if it happens to be doing work for you fighting national security threats abroad?

In the wake of the 2009 Green Movement in Iran, a group calling itself the Iranian Cyber Army emerged and began menacing Green Movement sympathizers at home and abroad. Hacking collectives had been active in Iran since the early 2000s, with groups like Ashiyane, Shabgard, and Simorgh cracking into websites for notoriety and, occasionally, profit. Beginning in the summer of 2009, however, politically motivated attacks on websites became increasingly common as a means to counter the Green Movement and create a climate of fear and suspicion. The Iranian Cyber Army hackers successfully defaced Twitter, Voice of America, the Chinese search engine Baidu, and opposition websites such as Radio Zamaneh, often emblazoning pages with their logo and leaving pro-government messages. (Recently, sophisticated attacks on the certificate authority systems that secure Internet traffic moving in and out of Iran were undertaken by an individual claiming to be a loner sympathetic to the regime, although no one can say for sure if the claim is true.)

The Iranian government has tacitly condoned the activities of the Iranian Cyber Army – even going so far as to applaud its efforts – while keeping itself one step removed from any formal endorsement or incorporation. When the Iranian Cyber Army launched a cyber attack on Voice of America websites and inserted an anti-American message, an Iranian official spokesman, Ali Saeedi Shahroudi, said that the U.S. could no longer claim that it was the "bellwether of software and cyber technology," and that the "hacking of a VOA homepage by the Iranian Cyber Army and leaving a message on the site for the U.S. secretary of state shows the power and capability of the [Islamic Revolution Guards] Corps in the cyber arena." In 2010, the leader of the Iranian Revolutionary Guard's Ali Ibn abi-Talib Corps, Ebrahim Jabbari, publicly claimed that his organization possessed the world's second-largest cyber army. Was he referring to the Iranian Cyber Army? In 2011, another Iranian Revolutionary Guard official, Brigadier General Gholamreza Jalali, said, "We welcome the presence of those hackers who are willing to work for the goals of the Islamic Republic with good will and revolutionary activities."

• • •

Quasi-national cyber armies like these are spreading, and spreading fast, for two fundamental reasons. First, the tools to engage in cyber attacks are now widely available and as simple to acquire as "download, point, and click." With such easy access, we have entered the age of do-it-yourself information warfare. A second factor, which reinforces and builds upon the first, is the growing pressure on governments and their armed forces to develop cyber warfare capabilities. While cyber warfare threats are often wildly exaggerated in order to win massive defence contracts, there is an undeniable arms race occurring in cyberspace, and the domain is being rapidly militarized. Governments around the world now see cyber security as an urgent priority. They are standing shoulder-to-shoulder with their armed forces on this issue, and the capacity to fight and win wars in cyberspace is now seen as an absolute necessity by authoritarian regimes and liberal democracies alike.

But not all countries follow the same playbook.

While the United States and other Western countries build cyber commands staffed by professionally trained military personnel, corrupt, autocratic, and authoritarian regimes follow a different path: exploiting the techniques and methods of the cyber-criminal underground, enlisting paramilitary hackers, and taking advantage of the vulnerabilities of the very systems their opponents depend on for mobilization and political action. They also target different adversaries, reflecting their own perception of what constitutes a national security threat: political opposition parties, independent media, bloggers and journalists, and the vast networks of civil society groups pressing for openness, democracy, and accountability.

For many years, global civil society networks saw the Internet and other forms of new media as powerful tools for their causes. They have gradually come to learn that these media can be controlled in ways that limit access to information and freedom of speech for citizens living behind national firewalls. Now, to those concerns must be added another, this time more ominous: cyberspace is becoming a dangerously weaponized and insecure environment. It is now a domain where human rights activists, opposition groups, and independent media can be trapped, harassed, and exploited, as much as they can be empowered. And there's another thing. On what basis can the West condemn, for instance, the Syrian Electronic Army or other quasi-state hacker groups for infiltrating the computers of opposition groups when we openly market offensive computer network attack products and services at Las Vegas–style trade shows?

10.

Fanning the Flames of Cyber Warfare

Eugene Kaspersky is the CEO of the Russian-based malware and cyber-security research laboratory that bears his name, Kaspersky Lab. An outspoken, controversial, and sometimes flamboyant figure in the computer security industry, Kaspersky attracted wide public attention in 2011 when his twenty-year-old son, Ivan, was kidnapped by people suspected of having ties to the Russian mafia. Ivan was quickly rescued by Russian security forces, and Kaspersky claimed no ransom had been paid ($4.5 million had been demanded). The incident led to considerable speculation. Russian secret forces do not typically intervene in kidnappings involving average citizens, but Kaspersky is no average Russian and many believe that he made a deal with authorities to gain the release of his son, which Kaspersky vehemently denies.

I first encountered Kaspersky at the London Conference on Cyberspace in November 2011. Organized by the British Foreign and Commonwealth Office, the conference was meant to be a major "rules of the road" meeting of great powers on the future of cyberspace and Kaspersky was among several high-profile speakers. The conference itself was poorly organized and produced no tangible results, but Kaspersky certainly made for good theatre.

Taking his turn at the podium, Kaspersky addressed the buttoned-down crowd. His tie askew, suit threadbare, and hair wild and unruly, he began with a finger-wagging admonition: "I am glad to see that people are finally taking this issue seriously. I have been warning about it for decades. If you had listened to me, and took me seriously, all those years ago."

After this stark beginning, Kaspersky segued into a more disturbing aspect of his lecture, a series of statements that left many in the assembled crowd squirming, me among them. Kaspersky is concerned about anonymity online, and that too many people are getting away with Internet crime because they can hide their tracks. He believes we need to institute the cyber equivalent of the passport or driver's licence. We do not allow people to drive cars without a licence, Kaspersky asserted, so why should we let them browse the Internet unchecked, unregulated? And then he went even further, suggesting that Russia should be regarded as a model for the rest of the world when it comes to Internet governance.

Russia? The model for the rest of the world!

Grumbling started at the back of the room and rippled forward. There were grimaces everywhere, especially among our British hosts, but there were also some vigorous nods of approval from law-and-order types, most of them sporting the dark blue suits and very short haircuts that are the uniform of the defence and intelligence community.

In fact, somewhat under the radar, Russia *has* indeed created a model for cyberspace governance for other autocratic regimes to follow. The Russian Internet, known locally as RUNET, accomplishes controls not through Internet censorship per se, which has been applied only selectively in the past and even then mostly around specific content categories, like homoerotic pornography. Instead, Russian authorities rely on more sophisticated, but also more brutal methods – intimidation, public discrediting, surveillance, and symbolic arrests – while also meddling in organized crime and employing patriotic hackers to muddy attribution. Unfortunately these tactics have proven attractive to a growing number of autocratic regimes looking to control information and digital activism. Russia, the model for the rest of the world? Maybe, if what we have in mind for the future of cyberspace is a *Blade Runner* dystopia.

Kaspersky raised another ominous possibility, telling *Sky News* at the London conference: "We are close, very close, to cyber terrorism. Perhaps already the criminals have sold their skills to the terrorists, and then . . . oh, God."

Cyber terrorism. The phrase points to a sense of heightened anxiety that has pervaded talk of cyber security since 9/11: the view that those hideous events represented a failure (at least in part) of Internet surveillance; that had control been tightened over digital communications the perpetrators might have been identified before they were able to execute their plan. But raising the spectre of cyber terrorism can also get a person discredited as a Cassandra.

In this respect, Kaspersky is the Russian equivalent of Richard A. Clarke, the former U.S. counterterrorism czar. Like Kaspersky, Clarke is famously outspoken, and both believe they were onto something long before anyone else. (Notably, Clarke warned his superiors about the threat of al-Qaeda targeting the United States prior to 9/11.) Like Kaspersky, he is often dismissed as an alarmist, seen by many as simply a rhetorical bomb-thrower. Clarke may not have been the first to employ the phrase "electronic Pearl Harbor" – it was John Deutch, the former CIA director, back in 1996 describing the prospect of terrorists using the Internet to launch a surprise attack – but he uses it liberally, as do many U.S. defence industry lobbyists. Indeed, the phrase continues to be repeated like a mantra in Washington. But if people like Clarke have been warning of a catastrophic electronic Pearl Harbor for decades, why hasn't it happened? Surely it is not for lack of people with grievances and access to computers?

The truth is that such extreme scenarios are unlikely for a number of reasons. The Internet (and cyberspace as a whole) is resilient precisely because governance over it is so distributed, and routing of network traffic across the Internet was designed from the outset to take multiple potential paths in the event of a failure of any one of them. The flip side, however, is that as cyberspace expands and embeds itself in more and more of everything we do, the chances of a cascading failure having catastrophic repercussions become considerable. In other words, it also seems unlikely that nothing bad will happen.

● ● ●

In June 2012, Kaspersky was back in the news, his company announcing that it had found a major cyber weapon called Flame. Better described as a tool of espionage than a weapon, Flame did not damage computers, but instead siphoned off massive volumes of information in a manner similar to GhostNet.

While technical experts pored over the data, some argued that there could be underlying political processes at work in the Flame revelations. Kaspersky's organization was given the Flame virus to examine by a Malaysia-based organization called IMPACT (the International Multilateral Partnership Against Cyber Threats), a public–private cyber security alliance set up in 2008 by the International Telecommunication Union (ITU). Founded in the late nineteenth century to enable governments to coordinate international postal and telegraphic traffic, the ITU is the world's oldest international organization, and its membership over the years has been almost entirely composed of state-owned telecommunications companies. (Some view it as a telecom cartel for just this reason. State-run telecommunications companies use the ITU to set long-distance telephone rates, a highly profitable source of government revenue.)

The ITU missed the boat on the Internet, however, which was developed largely outside the telecommunications sector and governed by engineers through an independent non-profit, the Internet Corporation for Assigned Names and Numbers (ICANN), under contract to the U.S. Department of Commerce. Over the past twenty years or so, as the Internet has grown enormously in importance, the ITU has tried to claw its way into Internet governance, a move at times fiercely resisted by those partial to the Internet's non-state system of governance. Nonetheless, the persistent threat posed by cyber weapons and warfare lend credibility to the involvement of ITU and IMPACT in cyber security and governance. Interestingly, Russia, China, and other governments fully support this involvement, seeing more UN- and ITU-based control as a way to legitimize their own vision of a territorially bounded system of global communications governance that aligns with national sovereignty. In 2011, for example, Russia, China, Tajikistan, and Uzbekistan proposed a "code of conduct" for cyberspace at the UN General Assembly, and Russia and China have been vocal proponents of a view of cyberspace governance that gives prominence to state controls over the Internet, and state organs power in the decision-making forums that set the rules of the road. Could the sharing of the Flame virus with Kaspersky's group by the ITU and IMPACT, and his trumpeting about finding a giant cyber weapon, be part of an overall campaign to lend support to the Russian and Chinese preferences for cyberspace governance?

The possible connections between Flame and another devastating cyber weapon, Stuxnet, fanned the flames of these suspicions. Stuxnet was discovered in 2010, and had been connected to devastating setbacks at Iranian nuclear enrichment facilities. In May 2012, when Kaspersky first made the announcement of the Flame discovery, he speculated that it belonged to the same family of malicious software as Stuxnet, and just about everyone who examined the case believed either the United States or Israel (or both acting together) were involved in its production. Only four days after Kaspersky's discovery of Flame, an explosive *New York Times* exclusive by journalist David E. Sanger all but confirmed those suspicions. Adding to the intrigue was the fact that the majority of the victims targeted by the Flame virus were in the Middle East, with most of them in Iran, and that later Kaspersky Lab claimed to have found an authorship link between a 2009 version of Stuxnet and Flame, a claim independently backed up by the security firm Symantec, and then by a supposed U.S. intelligence insider, who leaked the story to the *Washington Post*. As Roel Schouwenberg of Kaspersky Lab theorized: "I think this new discovery shows that the Stuxnet team used Flame code to effectively kick-start their project. I definitely think they are two separate teams, but we do believe they are two parallel projects commissioned by the same entities."

At the very moment that Russia, China, and their allies are pushing for greater international controls over cyberspace, their primary adversary, the U.S. and its ally Israel not only engage in but appear to tacitly acknowledge their responsibility for the world's first act of cyber sabotage against a critical infrastructure facility. As former NSA Director Michael Hayden remarked, "Somebody crossed the Rubicon." The age of cyber warfare is finally upon us.

11.

Stuxnet and the Argument for Clean War

News of Stuxnet first emerged in June 2010 when it was identified by a small Belarus security company, VirusBlokAda. Later, the German researcher Ralph Langner undertook a detailed "decoding" of the virus and helped determine that its target was the specific type of Siemens-produced equipment used at the Iranian Natanz nuclear facility. Speculation quickly grew that the Israelis and/or Americans were behind Stuxnet. Who else could disrupt Iranian nuclear enrichment plants with such stealth and precision? Either the Americans or Israelis, or both acting together, most assumed, and there was growing circumstantial evidence.

The Israelis are generally coy about their military prowess and secretive about their hardware (e.g., their nuclear weapons arsenal). Was it just a slip of the tongue at the retirement party for Lieutenant General Gabi Ashkenazi, the former head of the Israel Defense Forces, when celebrants appeared to claim Stuxnet as one of his major successes? (There was even an hilarious Israeli commercial done for a cable TV company showing what appears to be three bumbling Mossad agents undercover as hijab-wearing women in Iran blowing up a centrifuge after accidentally pressing a button on a Samsung tablet.) American officials also spilled some beans. In December 2010, Gary Samore, White House Coordinator for Arms Control and Weapons of Mass Destruction, Security and Arms Control, told the Foundation for Defense of

Democracies in Washington: "We're glad they are having trouble with their centrifuge machine and that we, the U.S. and its allies, are doing everything we can to make sure that we complicate matters for them."

The leaks and speculations on authorship obscure a more important point: the formidable weapon itself and the precedent it sets. A June 2012 *New York Times* article by David E. Sanger describes the planning and operational process behind the Stuxnet virus – how it began under President Bush as "Operation Olympic Games" (OOG), and was passed on to the Obama administration. Upon leaving office, Bush pressed Obama to continue the program, and Sanger describes Obama as being enthusiastic about it, even pushing forward with OOG despite errors in the coding that led to the virus spilling out beyond the Iranian targets to computers in other countries, and from there to the Belarus security firm.

The attack was planned and tested on a dummy Iranian nuclear enrichment plant, a fake target built from scratch in the United States. The *New York Times* reported that in early 2008 Siemens co-operated with the Idaho National Laboratory (part of the U.S. Department of Energy) to identify the vulnerabilities of Siemens computer controls used to operate industrial machinery around the world. From intelligence gathered by the Americans, it was known that Siemens equipment was being used in Iran's enrichment facilities. Around the same time, the Department of Homeland Security teamed up with the same Idaho lab to study a widely used Siemens control system known as PCS-7. The vulnerability of PCS-7 to cyber attack had been an open secret since Siemens and the Idaho National Lab outlined at a conference in July 2008 the kinds of manoeuvres that could exploit holes in its systems to meet a number of goals, including gaining remote control. Meanwhile, the Israelis started to experiment on an industrial sabotage protocol based on a mockup they had designed of Iran's enrichment program.

The code behind Stuxnet was far larger than a typical worm, considerably more detailed, and it contained some brilliantly crafted and highly suggestive elements, including clues as to Israel's direct involvement. Symantec researcher Liam Ó Murchú noted that his company had uncovered a reference to an obscure date in the worm's code: May 9, 1979, the day, shortly after the Iranian Revolution, when a prominent member of the Iranian Jewish community, Habib Elghanian, became the first Jew executed by the new Islamic government. Berlin-based security expert Felix Lindner then found that all manually written functions in Stuxnet's payload bore the time stamp "September 24, 2007," the day President Mahmoud Ahmadinejad first publicly questioned whether the Holocaust took place, during a speech at New York's Columbia University. Lindner found a file inside the code named Myrtus, and speculated this could be a reference to the Book of Esther, an Old Testament story where the Jews pre-empt a Persian plot to destroy them. It is hard to believe the Israelis would unwittingly leave such tell-tale signs of their involvement in Stuxnet; much more likely they show a deliberate intention to drop coy admissions of prowess.

A remarkable component of Stuxnet was its ability to cross "air-gapped" computing systems that are not actually connected to the Internet. In April 2012, the website Isssource.com, belonging to Industrial Safety and Security Source, published an article alleging that "former and serving U.S. intelligence officials" had said that an Iranian double agent working for Israel had inserted Stuxnet into the Iranian control systems using a corrupt memory stick. The article's author, former United Press International journalist Richard Sale, stated that the double agent was probably a member of the Iranian dissident group, the Mujahedeen-e Khalq (MEK), a shadowy organization with Israeli government connections that is believed to be behind the assassinations of key Iranian nuclear scientists.

Stuxnet was specifically designed to infect only certain types of supervisory control and data acquisition (SCADA) systems used for real-time data collection, and to control and monitor critical infrastructure – hydro-electrical facilities, power plants, nuclear enrichment systems, and so on. The programs used to control the physical components of SCADA systems are called programmable logic controllers (PLCs), and Stuxnet was developed in such a way as to target only two types of PLC models controlled by the Siemens Step 7 software – S7-315 and S7-417 – both of which are used in the Iranian nuclear centrifuges.

Stuxnet was designed to disable the centrifuges by inducing rapid fluctuations in the rotation speed of their motors. Unchecked, this would eventually cause them to blow apart, and one of the most remarkable aspects of the virus was a piece of deception created to confuse Iranian personnel monitoring the plants. Stuxnet secretly recorded what normal operations at the plant looked like, and then played these readings back to the plant operators (like a pre-recorded security tape) so that everything seemed to be in good order. While the operators were watching a normal set of operating results on their monitors, the centrifuges were actually spinning out of control. According to the *New York Times*, over and over again the Iranians sent teams of scientists down to the centrifuges with two-way radios, reporting back to the operators what they witnessed. They were utterly bewildered by the discrepancy between what they were seeing first-hand in the physical plant and what the monitors were reporting to the operators. Stuxnet was designed, as one insider put it, to make the Iranians "feel stupid."

• • •

While remarkably complex in some ways, Stuxnet is hardly extraordinary in others. Some analysts have described it as a

Frankenstein of existing cyber criminal tradecraft – bits and pieces of existing knowledge patched together to create a chimera. The analogy is apt and, just like the literary Frankenstein, the monster may come back to haunt its creators. The virus leaked out and infected computers in India, Indonesia, and even the U.S., a leak that occurred through an error in the code of a new variant of Stuxnet sent into the Natanz nuclear enrichment facility. This error allowed the Stuxnet worm to spread into an engineer's computer when it was hooked up to the centrifuges, and when he left the facility and connected his computer to the Internet the worm did not realize that its environment had changed. Stuxnet began spreading and replicating itself around the world. The Americans blamed the Israelis, who admitted nothing, but whoever was at fault, the toothpaste was out of the tube.

The real significance of Stuxnet lies not in its complexity, or in the political intrigue involved (including the calculated leaks), but in the threshold that it crossed: major governments taking at least implicit credit for a cyber weapon that sabotaged a critical infrastructure facility through computer coding. No longer was it possible to counter the Kasperskys and Clarkeses of the world with the retort that their fears were simply "theoretical." Stuxnet had demonstrated just what type of damage can be done with black code.

• • •

For some, Stuxnet represents a dangerous and highly unpredictable new form of conflict; for others, it taps into something far more attractive, the prospect of "clean" or "civilized" warfare: precise, surgical, virtual, and, most importantly, bloodless. "You're seeing an evolution of warfare that's really intriguing," argues Phil Lieberman, a security consultant and chief executive of Lieberman Software in Los Angeles, ". . . warfare where no one [dies]." The minister in charge of Britain's armed forces, Nick Harvey, echoes a similar sentiment: "[If] a government has arrived at the conclusion that it needs, out of its sense of national interest or national security, to deliver an effect against an adversary . . . arguably this [Stuxnet] is quite a civilized option."

The appeal of this argument is intuitive. If we can undertake acts of sabotage without killing or physically harming people, this does seem to represent progress, a new, gentler form of warfare. In this respect, the argument is the exact inverse of the neutron bomb debates of the 1970s and 1980s. The neutron bomb was an enhanced radiation weapon under development during the Carter and Reagan administrations that would kill people while leaving buildings and infrastructure intact, through a highly concentrated dispersal of radioactive material. (Soviet General Secretary Leonid Brezhnev memorably described it as a "capitalist bomb" because it would destroy people, but not property.) Stuxnet-type weapons, on the other hand, are more like something inspired by Unabomber Ted Kaczynski: they would target industrial-technological systems, but leave people alone.

The attraction of technology that allows one to believe in sanitary or "virtual war" has a long pedigree. Political scientist James Der Derian has spent considerable time turning over the argument, and believes that the appeal of high-tech means of fighting clean wars comes from it being "the closest we moderns have [come] to a *deus ex machina* swooping in from the skies to fix the dilemmas of world politics, virtually solving intractable political problems through technological means." But the "solutions" offered by virtual war mask the violence that invariably accompanies the use of high-impact technological weapons, and ignores the new problems and unforeseen consequences that arise. When high-tech weapons are marketed, available, and perceived as "clean," there are strong pressures to adopt military over diplomatic solutions in times of crisis. "When war becomes the first, rather than the last, means to achieve security in the new global disorder," says Der Derian, "what one technologically can do begins to dominate what one legally, ethically, and pragmatically should do." Meanwhile, the actual killing involved in warfare recedes into the background the more the application of force resembles a machine-like simulation or a computer game. "Virtuous war is anything *but* less destructive, deadly or bloody for those on the receiving end of the big technological stick."

Stuxnet-style attacks may seem like a higher order of sanitized conflict, but the Iranians undoubtedly do not feel that way. The question is, how will they react to Stuxnet? They may continue to develop and refine their own cyber warriors who will attack back with their own black code. In response to Stuxnet, Brigadier General Gholamreza Jalali, the head of Iran's Passive Defense Organization, said that the Iranian military was prepared "to fight our enemies [in] cyberspace and Internet warfare." Writing in the *Bulletin of the Atomic Scientists*, R. Scott Kemp argues, "Each new cyberattack becomes a template for other nations – or sub-national actors – looking for ideas. Stuxnet revealed numerous clever solutions that are now part of a standard playbook. A Stuxnet-like attack can now be replicated by merely competent programmers, instead of requiring innovative hacker elites. It is as if with every bomb dropped, the blueprints for how to make it immediately follow. In time, the strategic advantage will slowly fade and once-esoteric cyberweapons will slowly become weapons of the weak." And the Iranians' response may not come via cyberspace at all, but rather in a way that is as spectacular and grotesque as Stuxnet was stealthy and clean. We can now only wait and see.

Apart from unintended blowback, another dynamic bears closer scrutiny: the politically calculated revelations about Stuxnet being a U.S. and Israeli operation will most certainly fan arguments for the legitimacy – indeed, the urgency – of governments developing their own cyber warfare capabilities, or risk being left behind. Stuxnet did not start the cyber arms race, but it marks a major milestone and raises the bar considerably. And this is only the beginning. In October 2012, President Obama signed Presidential Policy Directive 20, authorizing the U.S. military to engage in cyber operations abroad to thwart cyber attacks on U.S. government and private networks. The directive establishes the "rules of engagement" to guide the operations. An unnamed senior administration official told the *Washington Post*: "What it does, really for the first time, is explicitly talk about how we will use cyber operations . . . Network defense is what you're doing inside your own networks . . . Cyber operations is stuff outside that space."

The world's most powerful state is now generally perceived as having been responsible for using computer code to successfully sabotage another country's critical infrastructure, and for ramping up offensive operations across the board. Not surprisingly, other countries are following suit. A 2011 study undertaken by James A. Lewis and Katrina Timlin of the Center for Strategic and International Studies – notably, done prior to the 2012 Stuxnet revelations – found that thirty-three states included cyber warfare in their military planning and organization, with twelve already having plans to establish cyber commands in their armed forces. Some, like India, boast about developing offensive cyber attack capabilities, while others are no doubt just being more discreet.

• • •

A few weeks after the Stuxnet revelations hit the news, there was a brief event that passed quickly through the news cycle but deserved more attention. Twitter went dark for a few minutes, leaving the global Twitterati at a complete loss. Speculation and rumours abounded. Was this the work of the hacktivist group Anonymous? The Iranian government? According to Twitter's company blog post, "Today's Turbulence Explained," the outage was due to a "cascading bug," a bit of malware "with an effect that isn't confined to a particular software element, but rather . . . 'cascades' into other elements as well." Tweeting in response, the father of cyberspace, science fiction author William Gibson (known on Twitter as @GreatDismal), laid out a simple but alarming hashtag: #andsoitbegins.

And so it begins, a series of cascading bugs reaching deeper and deeper into the infrastructure that surrounds us, bugs that are accidental, partly accidental, and accidental by design. In 2010, under Operation Network Raider, American authorities won thirty convictions and seized $143 million worth of counterfeit network computer equipment manufactured in China. (One man, Ehab Ashoor, bought counterfeit Cisco equipment from an online vendor located in China, and was intending to sell it to the U.S. Marine Corps for combat communications in Iraq.) In 2012, a year-long probe by the Senate Armed Services Committee found 1,800 cases of fake electronic components destined for American military equipment: 1 million bogus parts, mostly from discarded electronic waste being recycled in China. The report found the bogus parts in SH-60B helicopters, in C-130J and C-27J cargo planes, in the U.S. Navy's P-8A Poseidon plane. A July 2012 article in *Ars Technica* noted that "more than 500 days after Stuxnet the Siemens S7 has not been fixed." That same month, *Wired* reported on a Canadian company, RuggedCom, that makes equipment and software for critical industrial control systems. It had planted a backdoor (a means to remotely access a system) into one of their products, by design. The login credentials for the backdoor included a static username, "factory," assigned by the vendor that couldn't be changed by customers. The company-generated password was based on individual media access control (MAC) addresses for devices.

Researcher Justin W. Clarke (no relation to Richard Clarke) has shown how, by searching the Internet via the SHODAN search tool, anyone could discover MAC addresses for industrial control systems, and then employ a simple computer script he has engineered to log in to those industrial control systems. This is a far cry from the elaborate operational planning that went into Stuxnet: all that is involved is one person, one search, and one script, and the result is total access! Clarke quietly notified RuggedCom, which did nothing for months, leading him to go public with his discovery. "It is esoteric, it is obscure, but this equipment is everywhere," said Clarke, explaining his reasoning. "I was walking down the street and they had one of the traffic control cabinets that controls stop lights open and there was a RuggedCom switch, so while you and I may not see it, this is what's used in electric substations, in train control systems, in power plants and in the military. That's why I personally care about it so much." And as if this were not enough, the story ends on another menacing note: "RuggedCom, which is based in Canada, was recently purchased by the German conglomerate Siemens."

• • •

The evolution of human—computer interaction has taken many twists and turns over the decades but there is an undeniable trajectory. With the first giant mainframe computers – mechanical structures sprouting wires and vacuum tubes that took up entire rooms – one computer was shared by many people. Today almost half the world has access to several computing devices that they individually own and operate. At home I have a MacBook Air (more like a large mint wafer than a computer), a sleek Power Mac

G5 in my office, and the now omnipresent iPhone in my pocket. We are evolving into a species of ubiquitous computing, with tiny digital devices embedded in just about everything around us, much of it operating without any direct human intervention at all.

Eugene Kaspersky, Richard Clarke, and others may sound like broken records or self-serving fear-mongers, but there is no denying the evolving cyberspace ecosystem around us: we are building a digital edifice for the entire planet, and it sits above us like a house of cards. We are wrapping ourselves in expanding layers of digital instructions, protocols, and authentication mechanisms, some of them open, scrutinized, and regulated, but many closed, amorphous, and poised for abuse, buried in the black arts of espionage, intelligence gathering, and cyber and military affairs. Is it only a matter of time before the whole system collapses? "If one extrapolates into the future," Arthur Koestler once said with respect to the nuclear predicament, "the probability of disaster approaches statistical certainty." Is cyberspace any different?

Analogies to the Cold War and the logic of mutual assured destruction (MAD) come to mind. In those recent times, humans let their baser competitive instincts threaten civilization itself. But it didn't happen. And now? With critical infrastructure a vector for armed conflict and all of us interdependent to such a substantial degree, shouldn't the same perverse logic that restrained policymakers from dropping the atomic bomb restrain them from dropping cyber bombs? Cold War expert Fred Kaplan sums it up this way: "Cyberwar is very different from nuclear war: less destructive but also less tangible. Yet they're similar in one important way: It is illusory to talk about 'winning' either." Will our complex interdependence on shared communication systems reach a threshold of mutual assured crashes (MAC)? As theories, MAD, MAC, game theory, and so on all assume that human beings are rational and that, in the end, we will always act in our own best interests. But in evolutionary terms, humans are still very much connected to our animal instincts, to lizard brains that drive us towards our baser emotions and that occasionally interfere with the neat and tidy reasoning assumed by theories based on the rationality principle.

On June 8, 2012, there was a news update related to Flame. Researchers at Symantec noticed that the virus, which Kaspersky's team had now linked to the authors of Stuxnet, had begun silently removing itself from infected computers. They discovered the "suicide" commands by monitoring their own honeypot computer infected with the Flame virus, which was eliminated by the commands, and which left no trace after the job was done. Traces from the very machines it had once compromised, crawling back across the fibre-optic cables and radio waves to the black hole from which it emerged in the first place. Like a genie, back in the bottle, gone for the moment, but not extinguished.

12.

The Internet Is Officially Dead

San Francisco, February 2012. I arrive at the RSA Conference, one of the largest computer conferences and trade shows in the world. Held annually since 1991 in the San Francisco area, the event is managed by RSA Security, one of the leading cryptography companies in the United States. This year, the theme is "The Great Cipher Mightier Than the Sword." Ironic, I think to myself, in light of the recent rash of computer security breaches, including a major one on RSA itself that targeted its SecurID tokens, an authentication mechanism provided to thousands of employees of Fortune 500 companies to access networks remotely, and, as advertised, in a secure manner.

The June 2011 RSA breach hit the American security and defence industry particularly hard, and was one of several in 2011 that called into question the reliability of some of the most basic mechanisms of the Internet's infrastructure: certificates, authentication mechanisms, and encryption schemes all backed by name-brand, impressive corporations. These are the systems and companies we rely on to secure not only our desktops, but our entire way of life. Verizon, Cisco, RSA, and others, all now apparently the naked emperors of cyber security. By mid-2011 there also were breaches of Lockheed Martin, Epsilon, NASA, PBS, the European Space Agency, the FBI, and Citigroup. (I dubbed it "Breachfest 2011," and thought T-shirts should be made up with the list of victims emblazoned on the back like city stops on a rock tour.)

Despite the breach at RSA, the conference is still a must-stop on the international cyber security agenda. I grasp the size of the meeting as I walk through San Francisco's hilly streets from my hotel: thousands of geeks streaming along the sidewalks, growing in numbers as I approach the massive Moscone Center, a sprawling interconnected set of hangar-like buildings just off San Francisco's downtown core, and the setting for this year's conference.

I am featured on a panel on "Active Defence," the latest euphemism in security circles for striking back in cyberspace. Many U.S. companies and government agencies are so frustrated with their inability to deal with persistent attacks on their intellectual property and infrastructure, that they are exploring ways to go beyond defence, to reach out across borders and deal with the problem where it originates. With me are two retired U.S. generals, Kenneth Minihan and Michael Hayden. Minihan was the director of the National Security Agency under George Bush Senior, Hayden the director of both the CIA and the NSA. I meet both of them before the panel begins, General Hayden introducing himself with a slightly unsettling, piercing stare that evaporates as we exchange pleasantries. (I read once that if one were to contrive a caricature of the director of the world's leading spy agency he might look something like Hayden. Having finally met him, I can see why.) A bald, bespectacled man, Hayden looks straight out of central casting for James Bond villainy. Minihan is a less impressive figure physically, more like a retired uncle lounging in a fairway clubhouse. He and I share empty bromides about the weather. Volleying platitudes with people who once commanded the largest security and intelligence institutions on the planet, the apex of secrecy and power, is an unsettling experience. I look into Hayden's eyes for some hint of targeted killings or forced extraditions, but all I see is calm self-assurance.

As we sit down, my thoughts drift back to the vendor expo across the hall. A major trade fair exhibiting the latest devices, hardware, and over-the-top software from the computer security and defence industries, the expo is located in a facility the size of several football fields. Obviously, it is at least as important as the conference itself. The expo is a curious pastiche: monster trucks meet Gates; bikini-clad women hand out USB sticks to nerds in business suits; and shiny BMW motorcycles with precariously balanced laptops on their seats rotate on elevated platforms. Auctioneers of the sort one would see peddling stain removers at a county fair bark out sales pitches for the latest firewalls and antivirus software. Years ago at conferences like this, the trade-show themes were all about the "magic of connecting": connecting people in social networks; connecting computers to each other, and to the Internet. The theme of this year's bonanza is all about doing just the opposite: building borders, fences, and firewalls to keep unwanted intruders and hackers out. Slogans alluding to theft, espionage, and cyber attacks are emblazoned on posters and banners that hang from the ceiling over the scattered vendor booths. There is a party-like atmosphere, and a discomfiting feeling: "threats," it would appear, are something both to fear and to celebrate.

• • •

I walk by booths for companies with names like "AlienVault" and "CheckPoint," and stop to linger at the Narus booth. Headquartered in Sunnyvale, California, Narus Inc. was founded in 1997 by Israeli security specialists Ori Cohen and Stas Khirman, two men who had recognized a growing market for products that could sift through big data – that ever-expanding archive of our digital activities and selves – and collect and collate that information for law enforcement and intelligence-gathering purposes.

The company later moved to the United States, where Boeing would eventually snap it up, and Narus is now a wholly owned subsidiary of the massive defence contractor.

Narus was one of the first companies to offer deep packet inspection, the practice of diving into Internet data at critical chokepoints to precisely identify specific packets, protocols, and other bits of information. In 2006, Steve Bannerman, Narus's marketing vp, told *Wired* magazine, "Anything that comes through [an Internet protocol network] we can record . . . We can reconstruct emails along with attachments, see what web pages they clicked on; we can reconstruct their [Voice over Internet Protocol] calls." I first read about Narus's technology in a 2007 press release boasting about the company's ability to provide "real-time precision targeting, capturing and reconstruction of webmail traffic [including from] services such as Yahoo! Mail, MSN Hotmail, and Google Gmail," and that it "helps customers around the world like AT&T, Korea Telecom, KDDI, Telecom Egypt, Reliance India, Saudi Telecom, U.S. Cellular, Pakistan Telecom Authority." Not entirely a rogues' gallery, but nonetheless a disturbing list of state enterprises mostly belonging to countries with very mixed records in terms of human rights and judicial oversight.

As outlined in Chapter 2, in 2004, Narus received some very negative publicity when AT&T whistleblower Mark Klein revealed that the NSA was running an extralegal eavesdropping facility that spied on Americans using a Narus product, STA 6400. (As it turned out, the NSA facility in which Klein worked is located at 611 Folsom Street in San Francisco, only a block and a half from the vendor expo.) The revelations led to a lawsuit launched by EFF, and then to an amendment of the U.S. Foreign Intelligence Services Act. The new legislation didn't ban such practices; rather, it gave the companies who participated in it, like AT&T, retroactive immunity from prosecution. As a result, EFF's lawsuit was dismissed in 2009.

Narus re-emerged in the public spotlight during the 2011 Arab Spring when it was among several Western companies whose sales to regimes notorious for human rights violations were subject to increasingly close scrutiny by journalists, activists, and others. In Narus's case, its sales to Telecom Egypt of deep packet inspection and other monitoring systems led to concerns that Egypt's security service might have employed them to identify protesters' communications.

At the 2012 RSA Conference Narus was promoting its latest addition to its flagship NarusInsight traffic intelligence system: the CyberAnalytics application. The brochure partially read: "Narus provides real-time network traffic intelligence and analytics software that analyzes IP traffic and flow data to map the digital DNA (or behavior) of the network ...Through its patented analytics, Narus's carrier-class software detects patterns and anomalies that predict and identify security issues, misuse of network resources, suspicious or criminal activity, and other events that compromise the integrity of IP networks. NarusInsight protects and manages the largest IP networks around the world, and has been deployed with commercial and government installations on five continents."

Government installations on five continents? I asked myself, wondering what specific government installations in what countries this might refer to?

• • •

After thirty-three years of active service, Lieutenant General Kenneth A. Minihan retired from the U.S. Air Force on June 1, 1999. Towards the end of his celebrated career, he became the fourteenth director of the NSA and the Central Security Service, the most senior uniformed intelligence officer in the Department of Defense. (He also served as the director of the Defense Intelligence

Agency during the Clinton administration.) Retirement did not slow him down, and he did not move far from his prior places of employment, directing his efforts towards vigorously developing business opportunities in military and intelligence markets for the private sector. Minihan serves on numerous boards of directors – at the time of writing, Nexidia Inc., BAE Systems Inc., Arxan Technologies Inc., Neohapsis Inc., LGS Innovations LLC., VDIworks Inc., Circadence Corporation, GlassHouse Technologies Inc., ManTech International Corporation, The KEYW Holding Corporation, Fixmo Inc., Command Information Inc. – and the Paladin Capital Group, where he is managing director and focuses his attention on developing new investment opportunities for Paladin's Homeland Security Fund. According to its website, after 9/11 Paladin collaborated with Minihan to "discuss how private equity could play a vital role in developing and delivering effective products, technology and services for the homeland and global security sectors." The website goes on to say: "Paladin's investment thesis attracted luminaries in national security including former CIA Director, the Honorable James Woolsey, and former Secretary of the Army, the Honorable Togo West, Jr. and others." Minihan was also chairman of the Security Affairs Support Association (now known as the Intelligence and National Security Alliance, or INSA), the self-described "flagship operation for industry and government partnership to enhance intelligence business development." Today, the association represents about 150 member corporations, including major American defence contractors Booz Allen Hamilton, Boeing, BAE Systems, General Dynamics, Lockheed Martin, and Northrop Grumman.

Just before the panel discussion begins I lean over to Minihan: "How about that trade show across the hall?" I say.

"I know, isn't it fantastic!" Minihan replies with glee. "Most of these guys used to work for me. I walk down the hall and they say, 'Hi, General, I used to be on your team.'"

I feel slightly dismayed at the thought of a former director of the National Security Agency cheerleading a plethora of private sector spinoff companies, their representatives saluting him as he passes by. Welcome to the ever-growing cyber security industrial complex, a world where a rotating cast of characters moves in and out of national security agencies and the private sector companies that service them. Minihan is at the apex of this new complex.

As the lights dim in the hall and the spotlights blind me from the audience, the formal introductions of the panel begin. In that moment, I think to myself, *the Internet as we once knew it is officially dead.*

13.

A Zero Day No More

In the aftermath of the 2011 revolution that brought down Egypt's Hosni Mubarak, protesters burst into the building that housed the state security services and combed through thousands of documents left by the departing regime. Among the files listing paid informants, tortured confessions, and acts of secret manipulation was one rather exceptional document: a contract from an obscure British firm, Gamma Group International, selling what appeared to be special infiltration software to the Egyptian intelligence services. Gamma Group had given the Egyptian State Security Investigations Service (SSIS) a five-month free trial version of their system. Based on the trial testing, Egyptian authorities reported in a memo (also found in the cache) that "the system has a high-level penetration of any type of e-mail (Hotmail, Google, Yahoo)," and that it was successful "in breaking through personal accounts on Skype network, which is considered the most secure method of communication used by members of the elements of the harmful activity because it is encrypted." The memo discussed how the product enabled the "recording of voice and video chats; recording the movement of the target by using his computer and even recording him if the computer has a camera; full control of the target computer and the ability to copy anything on his computer." Also found were files on activists that contained transcripts of communications, including Skype conversations. The documents, which the protesters posted on the Internet, provided a glimpse into the black arts of the growing commercial market for offensive cyber warfare and surveillance technologies.

A more detailed peek – indeed, more like a strip show – was provided in December 2011 by the whistleblower organization, WikiLeaks. Working with a number of organizations (including Privacy International), the renegade outfit released what they dubbed the "Spy Files," a collection of restricted documents, brochures, and manuals from dozens of obscure companies. This type of information is never posted on the Web or circulated publicly; rather it is disseminated at closed-door industry conferences with exorbitant registration fees, meetings that are restricted to a narrow circle of intelligence, law enforcement, and defence agencies. Privacy International personnel managed to infiltrate this inner sanctum, gather up promotional materials, and use WikiLeaks to shed light on this underground but massively expanding industry. I wrote the foreword to the release of "Spy Files" for Privacy International (on whose advisory board I sit). "Not too long ago, Internet pundits mocked slow-footed authoritarian regimes and predicted their demise," I remarked. "Today, they are prime customers for the tradecraft of cyberspace controls."

Among the brochures in "Spy Files" is a PDF for a product, FinSpy, marketed by Gamma Group and described as a "Remote Monitoring and Infection Solution." The glossy brochure resembles something you might casually peruse in a dentist's office or perhaps at the Apple store, except that in Canada and many other countries, the product being advertised, if used by a consumer, would be in clear violation of the law. FinSpy breaks into and secretly monitors the computers of its unwitting targets. The brochure describes how FinSpy is a "field-proven Remote Monitoring Solution that enables Governments to face the current challenges of **monitoring Mobile and Security-Aware Targets** that regularly **change location**, use

encrypted and anonymous communication channels and **reside in foreign countries**" [bold and capitalization in the original]. Mobile and security-aware targets that change location? Reside in foreign countries? An apt description of the globally networked, popular insurrection that faced off against the Egyptian government in 2011. No wonder the product was sold there.

The brochure provides an overview of FinSpy that amounts to a laundry list of the seamy side of cyberspace, and worthy for that reason of some considerable scrutiny:

●●● *Bypassing of 40 regularly tested Anti-Virus systems.* Here, Gamma insists that its FinSpy product is so advanced that it escapes the detection of forty companies whose mission it is to protect customer computers from trojan horses, viruses, and computer worms . . . the very same type of trojan horse being manufactured by Gamma itself! FinSpy is a "zero-day" vulnerability; that is, its "signature" has not yet been discovered by antivirus companies like Norton and Symantec.

●●● *Covert communication with Headquarters.* Here, the company explains that it can infect targets and communicate back to those operating FinSpy without users knowing it. Ingenious, unethical, and, well, illegal – unless you are working for a secret agency whose activities are exempt from the law (which is, after all, a target group of Gamma).

●●● *Full Skype Monitoring (Calls, Chats, File Transfers, Video, Contact List).* No surprise here – Skype is used by many people who wrongly believe that it provides communications security – and the brochure gives an intriguing generic "use-case" of how FinSpy was used to monitor Skype: FinSpy was installed on several computer systems inside Internet cafés in order to monitor them for suspicious activity, especially Skype communications to foreign individuals. Using the webcam, pictures of the targets were taken while they were using the system.

Such details bring to mind stories that circulated in Egypt (where the product was sold to intelligence agencies). In June 2011, the *Wall Street Journal* reported that Egypt's security service listened in on Skype communications of young dissidents and that "an internal memo from the 'Electronic Penetration Department' even boasted it had intercepted one conversation in which an activist stressed the importance of using Skype 'because it cannot be penetrated online by any security device.'" That the means by which the Electronic Penetration Department did so was Gamma's FinSpy certainly adds missing colour to this "use-case"; that Egypt had something called an Electronic Penetration Department in the first place paints its government in a hue of blood red.

In addition to the contract that ransacking protesters stumbled across, the *Wall Street Journal* also reported on a "Top Secret" memo from Egypt's interior ministry. Dated January 1, 2011, it describes the five-month trial of a "high-level security system" produced by Gamma Group that succeeded in "hacking personal accounts on Skype." The *Journal* notes that "the system was being offered for €388,604 ($559,279), including the training of four officers to use it, by Gamma's Egyptian reseller, Modern Communication Systems."

These revelations underscored how lucrative the market for FinSpy-like products has become. They also confirmed fears that U.S.-based NGOs like Freedom House that were training Middle East activists to use tools like Skype to secure their communications were actually instilling a false sense of security when computers are commandeered by customized trojan horses like FinSpy. (To this day, I regularly encounter activists relying on Skype and other "secure communications" tools touted by "trainers." They should know better.)

Which brings us back to other features in Gamma's FinSpy brochure:

- *Recording of common communication like Email, Chats, and Voice-over-IP* and *Live Surveillance through Webcam and Microphone*. Surveilling webcams and microphones in real time was exactly what the Chinese hackers in the GhostNet espionage campaign achieved through Ghost RAT. Here the same capabilities are being professionally repackaged and marketed. As we noted when *Tracking GhostNet* was published: "We are used to having computers be our window to the world; it's time to get used to them looking back at us."

- *Country Tracing of Target*. This feature seems superfluous in light of FinSpy's other capabilities, but governments today face transnational networks of adversaries, and thus this is an important selling feature for agencies looking to penetrate and immobilize them. Perhaps to track them down and eliminate them.

- *Silent Extracting of Files from Hard Disk*. The silent part is interesting. Picture yourself working at your computer while files are being silently removed, without your knowledge, from your hard drive and down a fibre-optic cable.

- *'Process-based Keylogger' for faster analysis*. Somewhere, each one of your keystrokes is being recorded and analyzed — *as you are typing*. This means not just the words you are entering into a document, but each and every password you use for what you assume are secure websites and programs — like encryption and anonymizer tools – including, of course, the master password for your computer.

••• *Live Remote Forensics on Target System.* A standard reconnaissance task undertaken to provide a snapshot of a victim's computer layout, and thus vulnerabilities, so that they can be further exploited with other pieces of malware. Given FinSpy's other capabilities you have to wonder why this is necessary, but there you have it.

•••

The contracts, brochures, and obscure company names in the WikiLeaks collection – as endlessly fascinating as they may be – are but a glimpse into a vast labyrinth and arms race in cyberspace. It's wrong to describe this labyrinth as "underground" in the sense that today such attack tools are cheap and widely available, and attackers can mount their assaults at fibre-optic speed from anywhere on the planet to anywhere else; but "underground" is apt as they can also disguise their origins and mask responsibility, and, of course, the market for such products is dominated by shadowy security services. As Harvard's Joseph Nye, who has been assistant secretary of defense and chairman of the National Intelligence Council, argues, "The cyber domain of computers and related electronic activities is a complex man-made environment, and human adversaries are purposeful and intelligent. Mountains and oceans are hard to move, but portions of cyberspace can be turned on and off by throwing a switch. It is far cheaper and quicker to move electrons across the globe than to move large ships long distances." And it is far easier for the perpetrators to remain anonymous, hence the critiques of Eugene Kaspersky, Richard Clarke, and others, and the attacks on online anonymity itself.

War scholars have long understood that in an offence-dominant environment such as this, there is constant pressure to keep up. Fear and insecurity grow, threats lurk everywhere, and rash decisions lead to unexpected outcomes. For those in the defence and intelligence services industry this scenario represents an irresistibly attractive market opportunity. Some estimates value cyber-security military-industrial business at upwards of US$150 billion annually. Like Dwight Eisenhower's military-industrial complex before it, the cyber-security industrial complex is intimately connected to militarization processes in the West and, in particular, the U.S. major corporate giants that arose in the Cold War, such as Boeing and Northrop Grumman, are now positioning themselves to service the cyber security market. "We've identified cyber as one of our four key areas for growth for the next five years," says Tim McKnight, vice-president at Northrop's intelligence systems division. They have been joined by dozens of little-known niche outfits like Gamma, VUPEN, and Endgame. In an era of financial austerity, with so many industries squeezed by economic downturns, the growing cyber security sector represents a golden egg.

There are numerous good reasons for a thriving cyber security market. Dynamic networks need to constantly fend off malicious software, and the private sector generally produces the most efficient and agile responses. But when twinned with the growing desire among defence and intelligence agencies (and some companies) to monitor an ever-widening range of threats and to sometimes "strike back," the same market creates perverse dynamics. Securing cyberspace is only a part of the cyber security market: exploiting it, mining it for intelligence, and even propagating vulnerabilities that undermine and destabilize it are quickly becoming just as lucrative parts of the game.

In 2012, the satirical website the Onion published a news video calling Facebook a "massive online surveillance program run by the

CIA" and alleging "that Facebook has replaced almost every other CIA information gathering program." The video shows testimony from a fictional deputy director of the CIA, Christopher Sartinsky: "After years of secretly monitoring the public we were astounded so many people would willingly publicize where they live, their religious and political views, an alphabetized list of all their friends, personal email addresses, phone numbers, hundreds of photos of themselves, and even status updates about what they were doing moment to moment. It is truly a dream come true for the CIA."

Sometimes great satire is just too true. Of course, it truly *is* a dream come true for the CIA, and for the companies that sell social network monitoring products and services to the CIA (and other defence and intelligence agencies). When that market opportunity is combined with growing pressures on the private sector, including social network platforms themselves, to effectively police the Internet, accompanied by laws that relax independent oversight and judicial restraints, a very troubling mix of incentives emerges.

Consider Social360, a company that monitors social networks for other companies. It advertises a special "crisis-monitoring" service which aims to identify protester activities that might be threatening to companies tarnished by scandals. Although they don't publish to whom they sell their services, one can easily imagine this service being offered to oppressive regimes threatened by popular uprisings like those that arose during the Arab Spring.

Or consider the U.K.-based ThorpeGlen Company, a world leader in the design and development of mass data analysis and storage solutions for the security sector. On July 6, 2010, the company announced that it had created the "largest social network" in the world with more than 1.2 billion nodes. "A node on a social network is a person, piece of equipment or account," ThorpeGlen explains. "The network itself maps the linkages between nodes meaning that the flow of funds through bank accounts, the movement of people and materials within a production facility or the way in which people communicate with each other by e-mail or telephone can be visualized and analyzed." ThorpeGlen offers little explanation about how it acquires such node information, but in a 2008 web demo, its VP of global sales showed off one of the company's "lawful access" tools by mining a single week's worth of call data from 50 million users in Indonesia. The purpose was to find the dissident needle in the haystack. As the *London Review of Books* reported:

Of the 50 million subscribers ThorpeGlen processed, 48 million effectively belonged to 'one large group': they called one another, or their friends called friends of their friends; this set of people was dismissed. A further 400,000 subscriptions could be attributed to a few large 'nodes', with numbers belonging to call centres, shops and information services. The remaining groups ranged in size from two to 142 subscribers. Members of these groups only ever called each other – clear evidence of antisocial behaviour – and, in one extreme case, a group was identified in which all the subscribers only ever called a single number at the centre of the web. This section of the ThorpeGlen presentation ended with one word: 'WHY??'

"Why??" indeed. What does this analysis prove? Beneath the slick presentation, the demo suggests that ThorpeGlen had access to real user data in Indonesia, presumably shared with the company by cellphone and other telecommunications companies. One company, one case, one country. But doesn't this beg two questions: how many other ThorpeGlens are out there mining our social network data? And, how many countries are doing what Indonesia and Indonesian telecom companies presumably did in 2008: share users' data without their consent with a private company servicing law enforcement and intelligence?

In 2011, the German hacker collective, Chaos Computer Club (CCC) announced that it had discovered and examined a backdoor trojan horse made by the German company DigiTask as part of a "lawful interception" program to listen in on Internet-based communications. In Germany, courts have long allowed the use of backdoor programs to help law enforcement listen in on encrypted communications as part of legal wiretaps. However, the CCC alleged that the software went far beyond those permissible purposes, and claimed the trojan could be used to monitor Skype, Yahoo! Messenger, and MSN Messenger; log keystrokes made through Firefox, Internet Explorer, and other browsers; and take screen captures of desktops. The CCC wrote that the "State Trojan" violated German law because it could also upload and execute programs remotely. "This means, an 'upgrade path' from [lawful spyware] to the full State Trojan's functionality is built-in right from the start. Activation of the computer's hardware like a microphone or camera can be used for room surveillance. The government malware can, unchecked by a judge, load extensions by remote control, [and] use the Trojan for other functions, including but not limited to eavesdropping."

A German lawyer said that one of his clients was infected with the trojan while travelling through a German airport. After his client was arrested the lawyer contacted the CCC, which found the infection in the client's computer. WikiLeaks documents show that in 2008 German law enforcement was working with DigiTask to develop software that could intercept Skype phone calls. DigiTask stated that the program that the CCC found was probably a tracking program it had sold to Bavaria in 2007, and admitted that it sold similar spyware to governments throughout Europe.

The digital arms trade for products and services around "active defence" may end up causing serious instability and chaos. Frustrated by their inability to prevent constant penetrations of their networks through passive defensive measures, it is becoming increasingly legitimate for companies to take retaliatory measures. As the desire for such active defence strategies mounts, firms like CrowdStrike and Mandiant now openly go "on the hunt," distinguishing their services by contrasting them with those of mere "protection firms." "It's a lot more fun to fight the adversary than to guard against him," Mandiant company founder Kevin Mandia told NPR, citing another industry expert who says that "there are dozens, if not hundreds, of service providers doing similar things to Mandiant."

One extremely lucrative part of this market involves the sale of fresh "exploitations" or undiscovered computer vulnerabilities not yet detected by the antivirus industry, like Gamma's Zero Day. A 2012 *Forbes* magazine investigation acquired a price list of zero-day vulnerabilities, offering another peek inside this otherwise closed industry. Want a fresh exploit that will target Adobe? That will cost anywhere from $5,000 to $30,000. Mac OS X? $20,000 to $50,000. Android? $30,000 to $60,000. One exploit targeting Apple's iOS system was reportedly sold to a U.S. agency for $250,000.

The *Forbes* report profiles a Bangkok middleman, "The Grugq," who was set to earn over $1 million annually acting as a digital-age arms broker between those who engineer fresh exploitations and their purchasers, usually U.S. and European government agencies. Clearly, the burgeoning industry includes small obscure firms, lone actors, and industry giants like Northrop Grumman and Raytheon.

Of course, much of the industry is shrouded in the type of secrecy that accompanies defence, law enforcement, and intelligence agencies and their practices and markets. Entire segments of the cyber-security industrial complex operate in the shadows, reaping millions from ballooning "black budgets" that escape public scrutiny and independent oversight. An occasional leak here or there, dedicated investigative reporting, or a careless boast made by someone like "The Grugq" represent the only real chances the general public has to gain insight into this dark trade.

One of the few companies not afraid to speak out is the French-headquartered VUPEN Security, which came to prominence when its hackers won a 2012 contest sponsored by Google to see if anyone could find a vulnerability in its Chrome browser. The prize was $60,000, but in exchange for publicly disclosing the vulnerability the winner had to help Google engineers plug the holes. VUPEN surprised everyone by turning down the prize. "We wouldn't share this with Google for even $1 million," said the company's president. "We don't want to give them any knowledge that can help them in fixing this exploit or other similar exploits. We want to keep this for our customers."

VUPEN says it only sells to law enforcement agencies under a nondisclosure agreement, and only then to law enforcement agencies in the NATO, ANZUS, or ASEAN alliances. This sounds principled, but it should be noted that those alliances include such luminaries of human rights non-compliance as Albania, Bulgaria, Croatia, Hungary, Romania, Slovenia, Slovakia, Spain, Indonesia, Malaysia, Thailand, Brunei, Burma, Cambodia, Laos, and Vietnam. Nonetheless, in a glossy VUPEN brochure in the WikiLeaks "Spy Files" archive, the company boasts that it "provides its customers . . . reports about critical vulnerabilities up to 9 months in advance before any patches are released."

One of the other means by which researchers have tracked the growing black market industry is through job postings. In 2012, Mikko Hyppönen of the security firm F-Secure took notice of an increasing number of postings from large companies advertising for skill sets that included offensive exploitation capabilities. For example, a search by Hyppönen of the massive defence contractor SAIC's job database using the keywords "top secret/SCI" and "exploit" returned over 137 job postings. Intriguingly, a 2012 job posting at defence contractor Booz Allen Hamilton for a "target network analyst" looked to recruit someone who could "exploit development for personal computer and mobile device operating systems, including Android, BlackBerry, iPhone and iPad."

A 2011 Bloomberg News exclusive (based on anonymous sources) provides a detailed description of a service offered by one U.S. company, Endgame. It is worth quoting at length:

> People who have seen the company pitch its technology – and who asked not to be named because the presentations were private – say Endgame executives will bring up maps of airports, parliament buildings, and corporate offices. The executives then create a list of the computers running inside the facilities, including what software the computers run, and a menu of attacks that could work against those particular systems. Endgame weaponry comes customized by region – the Middle East, Russia, Latin America, and China – with manuals, testing software, and "demo instructions." There are even target packs for democratic countries in Europe and other U.S. allies. Maui (product names tend toward alluring warm-weather locales) is a package of 25 zero-day exploits that runs clients $2.5 million a year. The Cayman botnet-analytics package gets you access to a database of Internet addresses, organization names, and worm types for hundreds of millions of infected computers, and costs $1.5 million. A government or other entity could launch sophisticated attacks against just about any adversary anywhere in the world for a grand total of $6 million. Ease of use is a premium. It's cyber warfare in a box.

A person who used to work for the company, and requested anonymity, gave me a look inside that box. When I asked him about the type of product that Endgame might sell and how it might work for a customer, he went away for a few minutes, hammered on a keyboard at his computer, and produced a printout containing a long list of IP addresses of computers based in government ministries in Iran, all of which were "checking in" to a botnet he was carefully monitoring. Armed with this knowledge, he could have injected malware into those machines and had them all under his effective control. "That," he said, pointing to the printout, "is the type of information a client – let's say an adversary of Iran – would pay a lot to access on a regular basis."

• • •

One of the better snapshots of the cyber exploit and surveillance industry comes from a major trade show called the Intelligence Support Systems (ISS), something of a lightning rod for privacy activists. (Some of the WikiLeaks/Privacy International Spy Files were obtained at this trade show.) The ISS expo is restricted to defence, intelligence, and law enforcement agencies, but its public website provides summaries of the type of topics being discussed and products and services being marketed. The ISS World Middle East and North Africa expo, scheduled for March 2013 in Dubai, will feature the following panels and presenters: "Exploiting Computer and Mobile Vulnerabilities for Electronic Surveillance," Chaouki Bekrar, CEO and Director of Vulnerability Research, VUPEN; "Challenging the IP Interception Problem: Know your enemy, use the right weapon!", Murat Balaban, President, Inforcept Network; "Monitoring Social Networking Sites for Actionable Intelligence," Nanda Kumar, Director, Paladion Networks; "Identify 'Unknown' Suspects Using Unique Movement Patterns Derived from High Accuracy, Historical Mass Geo-Location of Wireless Devices," Bhavin Shah, VP Marketing and Business Development, Polaris Wireless – and, not to be outdone, this presentation from Gamma Group (of Egyptian "Electronic Penetration Department" fame): "Governmental IT Intrusion: Applied hacking techniques used by government agencies," MJM, Gamma Group.

The sponsors' page for the trade show reads like a rogues' gallery. Here are some highlights:

- trovicor: headquartered in Munich, Germany and with affiliate offices in Europe, Middle East, Asia-Pacific, trovicor services "Law Enforcement and Government Agencies in the security sector with deployments in more than 100 countries."

- Al Fahad Group: providing national security solutions ranging from "Interception, mediation, comprehensive protocol decoding including webmail and web 2.0 services; evidence processing, forensics, fraud detection, surveillance and cyber intelligence . . . [a]cross our operations in the Middle East, North Africa and Europe."

- Hacking Team: "Proven by more than 10 years of worldwide adoption and designed to fulfill LEAs and Security Agencies higher expectations, newly released version 8 'Da Vinci' gives you total control over endpoint devices."

- Polaris Wireless: "With commercial deployments in EMEA and APAC regions, our lawful and mass location intercept solutions are ideal for tracking known/unknown targets to within 50 meters including urban and indoor areas."

- Semptian Technologies: headquartered in Shenzhen, China, a cyber-monitoring expert in "providing the technical LI means to intercept Internet, PSTN fixed telephone and mobile phone networks . . . Semptian helps Law Enforcement

Agencies accomplish their missions such as criminal investigation, counter-terrorism, intelligence gathering and network security."

The ISS is unabashed about the type of trade that takes place under its auspices, and leaves no stone unturned in defence of its practices. Tatiana Lucas, ISS's world program director, for instance, wrote a letter to the editor of the *Wall Street Journal* taking issue with an article that exposed the trade fair and its implications for civil liberties. In a remarkably candid argument for greater commercial surveillance opportunities in the wake of the Arab Spring, Lucas said that criticism of the industry would hurt the U.S. economy, which would be left in the dust by others less shy about entering the market: "Based on our work with customers from around the globe, we expect that most countries outside the U.S. and Western Europe will begin to place intercept mandates on social networks, especially following the Arab Spring. This would give U.S. companies an opportunity to develop such tools and thus create jobs."

As one might expect, given its cloistered character, the political economy of this cyber exploit, data mining, and surveillance industry is woven through with former staffers of the very agencies it serves – thousands of replicas of former NSA director Kenneth Minihan. For example, the Israeli intelligence services elite Unit 8200, responsible for that country's advanced electronic warfare capabilities, has spawned numerous alumni who have gone on to create leading-edge companies in the cyber exploit and surveillance business. Many of them, like Gil Shwed, the CEO of Check Point Software Technologies, have become billionaires. Capitalizing on the cyber security boon, Check Point's shares have risen more than 70 percent over the past two years. "It's almost impossible to find a technology company in Israel without people from 8200, and in many cases the entrepreneur, the manager, or the person who had an idea for the project will be from 8200," says Yair Cohen, a former brigadier general who once commanded Unit 8200. In the United States, meanwhile, the NSA partners with "cleared" universities to train students in cyber operations for intelligence, military, and law enforcement jobs. Though run at the universities, the programs are secret to all but a select group of faculty and students who pass the necessary national security clearances. The training generally includes offensive orientations: "We're trying to create more of these, and yes they have to know some of the things that hackers know, they have to know a lot of other things too, which is why you really want a good university to create these people for you," an NSA staffer told reporters. Indeed, a rotating cast of characters from the spook world is reinforced by norms of secrecy across the public and private sectors, while providing opportunities for business inside government agencies.

• • •

While most of these products and services are manufactured or offered by North American and European companies, the market's greatest opportunities may lie in the global South and East, where there is a potent combination of exponential technological growth and connectivity and autocratic regimes looking to shore up hierarchical controls against digitally mobilized populations. Although the shroud of secrecy is often difficult to penetrate (an already secretive industry combined with autocratic regimes leaves little public accountability). Privacy International has identified at least thirty British companies that it believes have sold surveillance technologies to countries with shoddy human rights records – Syria, Iran, Yemen, Bahrain, et cetera – and it estimates the revenues of the global surveillance industry at $5 billion annually. In

August 2011, a French company, Amesys, sold deep packet inspection systems to the Gaddafi regime that were deployed by security services to monitor Libyan dissidents. The regime also purchased technology from China's ZTE, and from a South African company, VASTech, capable of tapping into international phone calls. When asked to justify its sales to a regime that was murdering its own citizens, a spokesperson for the company said it sells "only to governments that are internationally recognized by the United Nations and are not subject to international sanctions." Although the Gaddafi regime was finally ousted, and much of this cyber spying infrastructure shut down, insiders claim that the monitoring capabilities were quietly reactivated, and cellphones, emails, and chats are once again being systematically scrutinized.

In July 2011, the *Washington Post* reported on a U.S. Air Force contract solicitation for a surveillance system to be employed in Iraq, designed to intercept calls and messages in order "to assist in combating criminal organizations and insurgencies." It specified that the product must be capable of maintaining a database of "a comprehensive catalog of targets, associates and relationships . . . With mapping overlays, it should have the ability to locate targets being monitored and a warning alarm of less than 10 minutes if two or more targets come within a defined distance of each other," the *Post* reported. An Air Force spokesperson said that the technology is similar to that used by American federal and state law enforcement agencies, and that its use would be protected by Iraq's "stringent surveillance laws." A Human Rights Watch report disagreed, finding Iraq's "information crime laws" to be "part of a broad effort by authorities to suppress peaceful dissent by criminalizing legitimate information sharing and networking activities."

In 2012, an investigation undertaken by Swedish television producers uncovered a huge surveillance market in Central Asia being serviced by the Swedish Telecom giant TeliaSonera, which had allegedly enabled the governments of Belarus, Uzbekistan, Azerbaijan, Tajikistan, Georgia, and Kazakhstan to spy on journalists, union leaders, and members of the political opposition. One whistleblower told the producers, "The Arab Spring prompted the regimes to tighten their surveillance . . . There's no limit to how much wiretapping is done, none at all."

In October 2011, Bloomberg News provided a striking overview of the technologies used in Iran to quell dissent and create a climate of fear and self-censorship. As elsewhere, apprehended activists were routinely presented with transcripts of their mobile phone calls, emails, and text messages. After examining more than 100 documents and conducting dozens of interviews with technicians and managers who worked on the systems, Bloomberg concluded that the technology was provided to Iranian authorities by Stockholm-based Ericsson, Creativity Software of the United Kingdom, and Dublin-based AdaptiveMobile. Ericsson had pitched a sophisticated tracking system to the Iranian mobile operator MCI, which it said could assist law enforcement to track users and archive locations for later analysis. Nokia Siemens Networks faced an international "No to Nokia" boycott, EU Parliamentary hearings, and a lawsuit filed in U.S. courts (but eventually dismissed) by relatives of imprisoned Iranians for selling its communications intercept products to Iranian law enforcement. The Bloomberg story quotes an imprisoned activist, Mansoureh Shojaee, who was shown transcripts of her own communications while being interrogated in Tehran's notorious Evin Prison: "My mobile phone was my enemy, my laptop was my enemy, my landline was my enemy," she said.

On January 15, 2013, Citizen Lab researchers used a combination of technical interrogation methods to scan the Internet to look for signature evidence of censorship and surveillance devices associated with the American company, Blue Coat Systems. While our investigation was not exhaustive, what we did find raised alarm bells. We identified 61 Blue Coat ProxySG (designed for filtering and censorship) and Blue Coat PacketShaper devices (used for surveillance) on public or government networks in countries with a history of human rights abuses, surveillance, and censorship. Although both of the products have legitimate uses, their deployment in such contexts should be cause for everyone's concern.

Bloomberg News and the *Wall Street Journal* have sections on their websites – "Wired for Repression" and "Censorship Inc.", respectively – dedicated to the rapidly expanding cyber security industrial complex. Tools to track cellphones, deep packet inspection, social network analysis, and computer network attack and exploitation are being developed by firms the world over and sold to regimes seeking to isolate and arrest dissidents and activists, and to strengthen strangleholds over communications within their borders.

• • •

Can this market be regulated? Would export restrictions of the sort placed on advanced munitions make a difference?

In September 2011, the EU Parliament passed a resolution that bans the export of information technology systems that can be used "in connection with a violation of human rights, democratic principles or freedom of speech ... by using interception technologies and digital data transfer devices for monitoring mobile phones and text messages and targeted surveillance of Internet use." A strong and principled position, but far from flawless.

The same deep packet inspection systems used to spy on Libyan or Bahraini activists have legitimate purposes, like controlling against spam and other malicious flows of communication, but it is highly debatable that these functions are separated out by regimes and agencies not transparent about how they employ them. American political scientist Milton Mueller has argued, "The problem with this approach is that information technology, unlike bombs or tanks, is fundamentally multi-purpose in nature. You cannot isolate 'bad' information technology in order to control bad uses. There is no technical difference between the devices and services for digital surveillance used by the Chinese and Iranian governments and those used by the American, Canadian, French or British governments. The same capabilities inhere in all of them."

Moreover, attempts to regulate do not get at the root of the problem – the demand for such technologies. And this brings us back to the responsibilities the West has in driving the cyber-security industrial complex forward in the first place. Since 9/11, and with unrelenting momentum, liberal democracies have moved towards the normalization of what Yale University law professor Jack Balkin calls "the national-surveillance state." Whereas once it was fashionable to argue that the Internet would bring about the end of authoritarianism, how cyberspace is now being used and, more specifically, the new and emergent tools and tradecraft of surveillance and targeted attacks, suggest just the opposite. Summarizing Balkin's concerns, a 2012 *New Yorker* essay reported that since 9/11 the U.S. has witnessed "the emergence of a vast security bureaucracy in which at least two and a half million people hold confidential, secret, or top secret clearances; huge expenditures on electronic monitoring, along with a reinterpretation of the law in order to sanction it; and corporate partnerships with the government that have transformed the counterterrorism industry into a powerful lobbying force." More or less the same tendencies towards illiberal policies can be found in countries like Canada, across Europe, and parts of Asia. As long as law enforcement and intelligence agencies in such countries continue to drive demand, the cyber-security industrial complex will continue to expand worldwide and the surveillance society will be a fact of life at home and abroad.

• • •

July 2012. Bahrain's already restrictive media controls are ratcheted up. Bloggers and activists are increasingly at risk, many of them arrested and sentenced to lengthy prison terms for criticizing the regime or using social media to organize opposition campaigns. Once again Bahraini activists report experiencing targeted phishing and malware attacks, some of genuine sophistication, and dissidents arrested by authorities are presented with transcripts of their own text messages during interrogations.

The Citizen Lab's Morgan Marquis-Boire is contacted by Vernon Silver, a Bloomberg investigative journalist who has received what he believes is a high-grade trojan horse that has been menacing Bahrain's Net dissidents. Marquis-Boire contacts the Lab's security analyst Seth Hardy, a man who spent many years in the antivirus industry reverse-engineering sophisticated malicious software. What he sees is unprecedented in its complexity, its cloaking features "several orders of magnitude better than anything I have ever seen," says Hardy. This produces palpable excitement in the Lab, and Marquis-Boire seeks me out on a secure channel of communications. He describes the malware's sophisticated features, especially the way it masks itself within a computer, and then says that he was able to unravel a signature that connects the malware to its manufacturer. "We know who made it," Marquis-Boire says. "We have proof that it is Gamma's FinSpy."

A zero day no more.

14.

Anonymous: Expect Us

> But at this terminal point, where the automatic process is
> on the verge of creating a whole race of acquiescent and
> obedient human automatons, the forces of life have begun,
> sometimes stealthily, sometimes ostentatiously, to re-assert
> themselves in the only form that is left them: an explosive
> affirmation of the primal energies of the organism.
>
> — Lewis Mumford, *The Pentagon of Power*

June 3, 2011. A video is posted on YouTube from those
outlaws of the Net: Anonymous. It is a still image of a now-classic
Anonymous poster: blue and black shading, a frightening looking
lineup of men in suits topped with question marks where their
heads should be. Hovering above is the overlord Guy Fawkes, brim
down, covering his gaze in menacing fashion. Underneath, in large
letters, is the caption "Expect Us." A computerized voice-over,
backed by a pulsating symphonic score, is addressed directly to the
world's largest and most formidable military alliance: "Good eve-
ning, NATO. We are Anonymous. It has come to our attention that
a NATO draft report has classified Anonymous a potential threat to
member states' security, and that you seek retaliation against us."

The voice-over then offers up a critique of the NATO draft document, alludes to recent Anonymous hacks of the private American
security company HBGary, and in short, clipped sentences makes
its threatening concluding argument:

Anonymous is not simply 'a group of super hackers.' Anonymous is the embodiment of freedom on the Web. We exist as a result of the Internet, and humanity itself. This frightens you. It only seems natural that it would. Governments, corporations, and militaries know how to control individuals. It frustrates you that you do not control us. We have moved to a world where our freedom is in our own hands. We owe you nothing for it. We stand for freedom for every person around the world. You stand in our way. We hope you come to see that your attempts to censor and control our existence are futile. But if this is not the case, if you continue to object to our freedoms, we shall not relent. We do not fear your tyranny. You cannot win a battle against an entity you do not understand. You can take down our networks, arrest every single one of us that you can backtrace, read every bit of data ever shared from computer to computer for the rest of this age, and you will still lose. So come at me, bro. You can retaliate against us in any manner you choose. Lock down the Web. Throw us in prison. Take it all away from us. Anonymous will live on. We are Anonymous. We are legion. We do not forgive. We do not forget. Expect us.

Less than a year later, in an Anonymous signature moment, the movement posts an intercepted recording of a conference call between the FBI and Scotland Yard. The topic of the conference call? Anonymous itself. The call starts out with a few casual exchanges – jokes and observations about the weather – before moving on to the topic of rounding up people suspected of links to Anonymous, little doubt those behind the intercepted recording itself.

YouTube videos and other online statements such as these have become part of the Anonymous brand: brazen, irreverent, and almost always juvenile. Their videos typically include an ironic mixture of do-it-yourself editing tricks, silly Internet memes, pop culture allusions, and X-rated vulgarity topped off with petty anarchism. Part of me enjoys the videos, particularly those like the one about NATO that take a swipe at the defence and intelligence establishment. But another part of me sees them in a more troubling light. I am not so interested in the "who?" of Anonymous but in what their fight represents: resistance and rage against a state-security lockdown of the Internet. With each new video, each new Anonymous breach, a little part of me shudders, and I think of the other shoe dropping. At what point will taunts directed at the CIA, or NSA, or FBI finally wake up the bear? How long will they tolerate such open challenges to their power and legitimacy? And when they do lose patience – and no doubt they will – Anonymous will play right into their desire to do away with anonymity online altogether.

Part of me also thinks of the strategic benefit of Anonymous to those in power. As a child of the Watergate era – and an admirer of conspiratorial 1970s films about the dark forces pulling strings behind the scenes of government: *Three Days of Condor, The Parallax View, All the President's Men* – I often wonder just how many of the attacks for which Anonymous takes credit are actually the work of the very intelligence agencies being targeted? As Harvard law professor Jonathan Zittrain puts it, "Anonymous could be anyone, it could be the government, we don't know." Indeed, it would not be difficult to imagine a clandestine operative working for the Americans or the British or another government seeding an AnonOp, the name given for operations undertaken by Anonymous. How about one that meddles with an adversary by giving them a taste of their own medicine? Could this have been what was behind Anonymous's March 2012 sudden preoccupation with China? As The Who's "Baba O'Riley" played over and over again on defaced websites containing links to circumvent Internet censorship, an Anonymous screed warned the Chinese government that it is "not infallible, today websites are hacked, tomorrow it will be your vile regime that will fall." At that very moment, several Chinese companies experienced data breaches, the stolen data posted to file-sharing sites. A taste of China's own medicine?

• • •

By most accounts Anonymous's origins stem from the 4chan message board, one of the many dark alleys of the Internet, like a Lower East Side of cyberspace where every delinquent, offbeat, perverted taunt is not only tolerated but applauded. Anonymous spilled out of 4chan as a social movement in 2008, sparked when a decision taken by the Church of Scientology was viewed as a step too far across the breach of Internet morality. The Church sought to quash embarrassing online videos circulating across the Internet in typical meme-like fashion of a giddy Tom Cruise proclaiming his adherence to Scientology. A group calling itself Anonymous appeared, donned the now-familiar Guy Fawkes masks, and then started taunting the Church, both across the Internet and on the streets of cities throughout North America and Europe.

The movement remained obscure until the WikiLeaks saga and then the Arab Spring, when it unleashed a spree of overtly political AnonOps targetting what the amorphous mob claimed were foes of Internet freedom. It began with defacing and breaching attacks against websites and servers of a bewildering and some-times confusing array: the Tunisian, Egyptian, Zimbabwean, Malaysian, Libyan, and other governments; private companies like Sony, accused of censorship in the guise of protecting its intellec-tual property; financial services companies like Mastercard, PayPal, and Visa (for boycotting donations made to WikiLeaks); and the CIA, NSA, FBI, U.S. Department of Justice, and police forces around the world. Twitter accounts with the prefix "Anon" proliferated, and at one point in the fall of 2011, it appeared that Anonymous and the Occupy movement would consolidate into a powerful social force threatening the elites of the industrialized world – a more mature, digitally empowered next-generation version of the 1990s anti-globalization movement.

But then a series of dragnet-style arrests took place. Beginning in July 2011, and coordinated across the U.S., U.K., and the Netherlands, twenty people were detained. This was followed in February 2012 with Operation Unmask, coordinated by law enforcement agencies in Chile, Argentina, Colombia, and Spain, and resulting in the arrest of twenty-five people, followed by another wave of arrests in March 2012. Confirming at least some of my suspicions, the FBI had quietly arrested and then turned over a prominent member of LulzSec in 2011, who helped secure the arrests for the police. Nicknamed "Sabu," Hector Xavier Monsegur was charged with twelve counts of criminal conspiracy, and faced a maximum sentence of 124 years in prison. He secretly pleaded guilty and agreed to operate as an informer for the FBI to build cases for future arrests. The arrests (and later revelations about the turning of Sabu) dropped a poison pill into the networked well of Anonymous, and as 2012 rolled onward the number of AnonOps began to decline.

In analyzing Anonymous it is tempting to focus on salacious details: Who are the members? The ringleaders? What drives them to do what they do? The general impression might be white, nerdy, middle-class teens, a neat template for the Hollywood image of the "hacker." Some do, in fact, fit this image: for example, Ryan Cleary, a nineteen-year-old member of LulzSec living at home with his parents, was arrested in June 2011 during the Scotland Yard and FBI probe. His counsel told the court that Cleary suffered from both Asperger's syndrome and agoraphobia. He was subsequently given bail, under the condition that he stay off the Internet. But twenty-eight-year-old Sabu, who is of Puerto Rican descent and an unemployed foster parent of two children, clearly does not. Nor did the twenty-five individuals, mainly Latin Americans, arrested as part of Operation Unmask. The truth is, anyone can become part of Anonymous – that's the point, and there will be future Operation Unmasks and future iterations of Anonymous: Expect it.

• • •

Anonymous's methods fall into two general categories: breaches of computer systems and DDOS attacks. Breaches of computer systems are undertaken either by using malicious code that exploits a vulnerability in a server, or by fooling someone into giving you access to data, a technique known as "social engineering." Anonymous's breaches are typically followed by the exfiltration of data from targeted victims, and the publication of private, embarrassing, and/or incriminating information, like the massive Stratfor breach, which led to Anonymous turning over tens of thousands of proprietary company emails and email credentials of Stratfor subscribers to WikiLeaks. (At the time, WikiLeaks noted: "The material shows how a private intelligence agency works, and how they target individuals for their corporate and government clients.") Typically these are posted to sites like Pastebin, a resource primarily used to share bits of computer code but repurposed for Anonymous-style disclosures of data and announcements of successful attacks.

Most Anonymous DDOS attacks employ a crowd-sourced piling-on against targeted websites, using their preferred Low Orbit Ion Cannon (LOIC), a DDOS attack application that sympathetic users are encouraged to download and employ against a chosen victim. When used in numbers (i.e., in a "distributed" way), the LOIC makes repeated requests to servers from so many users that the servers are overwhelmed, taking them offline for a period of time. In cases where financial firms and retailers are involved, the DDOS attacks can result in significant losses of revenue. In 2012, Neustar, an Internet analytics company, surveyed IT professionals from twenty-six different industries to understand what was at stake during a DDOS attack. Over half of the companies surveyed reported that a DDOS outage would cause substantial financial damage, with 82 percent of financial firms estimating losses at more than $10,000 per hour, and 67 percent of retailers at $100,000 per hour. Beyond financial losses, companies also reported fears of damage to brand reputation and customer service experiences.

The DDOS attacks employed by Anonymous, though higher in profile than many others in recent years, are certainly not new. DDOS attacks have been going on for decades on the Internet, mostly launched by cyber criminals for extortion or other nefarious purposes. I first heard about politically motivated DDOS attacks in 1998, with reference to those organized by the New York–based hacker and artist collective, the Electronic Disturbance Theater (EDT). Led by the charismatic Ricardo Dominguez (now a professor of media studies), the EDT organized DDOS attacks against Mexican government servers in support of the Zapatista movement for autonomy in the Mexican province of Chiapas. Dominguez and his group openly advocated widespread participation in the DDOS attacks not only against Mexico but also against the U.S. Defense Department and other targets seen as sympathetic to Mexico. The attacks combined art and digital activism, loading up their DDOS tool with requests for non-existent content and sending these requests to Mexican government servers. When network administrators looked over their logs after the DDOS attacks, they saw results like "Ana Hernandez: Not Found," she being one of many Chiapan dead. The computers used by Dominguez and his group became the object of a counterattack by American law enforcement, one of the first active defence initiatives that are now so prevalent.

(At the time of the Zapatista cyber resistance, I was still formulating ideas for the collaborative research effort that would later become the Citizen Lab. Also living in Toronto at the time was Oxblood Ruffin, the self-appointed "foreign affairs minister" of one of the world's oldest, most respected, and principled hacker collectives, The Cult of the Dead Cow, or cDc. Oxblood and others were forming a politically charged subgroup of cDc called Hacktivismo, and we had discussions about the limits of acceptable political action online and the philosophy that would underpin Hacktivismo and the Citizen Lab. We agreed that DDOS attacks were unjustifiable except in extreme circumstances and that they were contrary to human rights because they infringe upon free speech. We still share that view.)

Some have tried to downplay DDOS attacks, even legitimize them. The Internet pundit Evgeny Morozov, for instance, has likened them to picket lines and sit-ins, the electronic equivalent of civil disobedience. But even Morozov recognizes the analogy only goes so far. Picket lines, sit-ins, and civil disobedience, as traditionally understood, all entail accepting the possibility (even the probability) of considerable personal consequences in the name of some higher moral good. DDOS attacks, on the other hand, can be carried out anonymously, usually without participants accepting legal consequences, and they involve little effort or cost. They are more akin to armchair activism, which raises the question: "Can an act of disruption undertaken without getting out of your seat and that has no likely legal repercussions be considered a legitimate form of civil disobedience?" (Such activism, however, can have serious unintended consequences, generally not for the armchair activists but for others. For instance, after Anonymous's Operation

Tunisia – largely mounted by hacktivists in North America and Europe – it was *Tunisian* bloggers and activists who were the ones arrested and had their computers confiscated).

More importantly, with the tools to cause havoc so cheap and readily available, and the consequences so potentially low, is it wise to actually encourage DDOS attacks as a form of political protest? Yale University's Yochai Benkler thinks so: "Except in extreme cases akin to the real-world burning of cars and smashing of windows (e.g., had PayPal's payment systems been disrupted and customers lost money, rather than the company's homepage being unavailable), they should simply be absorbed as part of the normal flow of the Internet. When addressed, these actions should be treated as a disruption to the quality of life, similar to graffiti." And yet, it is not unrealistic to imagine a kind of mass vigilantism in which any person with an axe to grind and a cheap laptop could seriously pollute, even bring to a halt, the free exchange of ideas through the global Internet. Don't like what someone says online? Blast them offline with a Low Orbit Ion Cannon. I cannot imagine any serious advocate of liberal democracy welcoming that prospect and, for that reason, I don't see this form of political action as justifiable. At the same time, it is not something that should be treated as a national security threat.

Putting aside the "who" and the "how" of Anonymous, the deeper question is why? Why has Anonymous erupted now, and what does this phenomenon represent? One of the few to study this question in depth is McGill University anthropologist Gabriella Coleman (who admits that after years of analyzing Anonymous she still has trouble answering the question, "Who is Anonymous?"). Anonymous is not an organization, Coleman believes, it's a name adopted by a range of groups to describe a wide array of actions linked in spirit and that share a certain disdain for authority. The few figureheads that have been arrested are not, for Coleman, emblematic of what Anonymous as a social movement represents: "They have tapped into a deep disenchantment with the status quo as concerns censorship, privacy, and surveillance . . . and they dramatize the importance of anonymity and privacy in an era when both are rapidly eroding." For Coleman the central, most interesting, point is the deep well from which Anonymous has emerged: "Irreverent dissent on the Internet is not going to go away with Anonymous," she asserts.

Is Anonymous a spontaneous reaction to growing controls over cyberspace, a crude affirmation of the human desire for freedom, and a reflexive, almost unconscious, self-protective mechanism against stifling constraints? Is it a kind of autoimmune response by cyberspace itself? A rage against the machine? If so, will it end up being counter-productive: the rage provoking, even infuriating the machine?

• • •

What is a hacker? For many the term conjures up images of a young, hoodie-wearing criminal bent over a keyboard, connecting remotely to an unwitting person's computer, siphoning off money from a bank account in some far-off jurisdiction or engaged in untoward cyberspace activities meant to upset the order of things or simply to embarrass some powerful person or entity, somewhere. Like Anonymous itself, rarely, if ever, is computer hacking considered benign, let alone useful. In the FBI's intelligence assessment of Anonymous a *hacker* is defined as someone who "conducts cyber intrusions to obtain trade secrets, financial information, or sensitive information," while a *hacktivist* is "someone who conducts a cybercrime to communicate a politically or socially motivated message." Either way, according to the FBI, to hack is to break the law.

It was not always thus: indeed computer hacking once had

positive connotations. Its origins date back to the late 1950s at the Massachusetts Institute of Technology (MIT), first surfacing among the engineers of MIT's Tech Model Railroad Club, who playfully referred to themselves as hackers. When the first mainframe computers were introduced at MIT soon thereafter, the hackers turned to fiddling with the machines in the same way as they did with trains. The term gradually embedded itself into the MIT computer science and engineering community by way of describing a curiosity about technology. A hacker was someone who did not accept technology at face value, and who experimented with technical systems, exploring their limits and possibilities: that is, a hacker opened up technical systems and explored their inner workings.

This original positive idea of hacking is what I had in mind in setting out to create a research hothouse that would bring together computer and social scientists. Hacktivism by my definition is the combination of social and political activism with that original hacker ethic, and this captures the gist of what I was hoping for in founding the Citizen Lab. Oriented around a specific set of values that would inform our research, as I saw it (and still do) hacktivism has a lot in common with a philosophical tradition stretching back to the ecological holism of Harold Innis, the pragmatism and experimentalism of William James and John Dewey, and the yearning for a return to a polytechnic culture of the early Renaissance articulated by Lewis Mumford. These thinkers all shared a particular view of technology as something that should be seen not as a thing or product, but as a *technic*, a craft, that was inherently political and essential to a healthy, democratic, and public life. Just as Mumford saw Leonardo da Vinci as the paradigmatic proto-citizen of a polytechnic society, I saw him as a prototypical hacktivist: interdisciplinary and experimentalist.

To my ongoing frustration, the term *hacker* has been corrupted and redefined, in part because of the actions of some hackers themselves. Irreverence towards authority has always been an element of the hacker spirit and ethic, and those who defined themselves as hackers would regularly find ways to step over acceptable limits, mostly for humorous ends. MIT Museum Hack archivist Brian Leibowitz notes that in the 1960s students on campus began to use the word as a noun to describe a great prank, and by the late 1960s the meaning included activities that "tested limits of skill, imagination, and wits." By the mid-1980s, the term was primarily being used at MIT to describe "pranks" and "unapproved exploring" of parts of the Institute or inaccessible places on campus.

Over time hacking came to connote a wide range of often extreme methods and ends. Steven Levy, author of *Hackers*, points out that "the word now has two branches, one used among computer programmers and the one used in the media." But few self-identified hackers remain faithful to the original spirit and ethic that first attracted people like Oxblood Ruffin and me; and, worse, today "hacker" and "hacking" are almost entirely synonymous with criminal acts, one or the other word invariably emblazoned in headlines each time Anonymous strikes or a data breach occurs. That is, "computer hacking" is used unquestioningly to describe anyone who breaks the law or causes a ruckus in cyberspace.

The association of the term with criminality is not just a semantic issue; it represents a much larger delegitimization of the underlying philosophy of experimentation at the heart of the hacker ethic. And herein lies an enormously important paradox, one that sits at the heart of our technologically saturated world: we have created a communications environment that is utterly dependent on existing (and emerging) technologies, and yet, at the same time, we are actively discouraging experimentation with, and an understanding of, these technologies. Never before in human history have we been so constantly plugged in and utterly connected. We are immersed in cyberspace, surrounded by technical systems embedded in just about everything we do, and to an ever-increasing extent they govern what we do, or, more accurately what we can and cannot do. In this context, the numerous and increasingly severe restrictions on what we are allowed to do with and within cyberspace are alarming.

• • •

The appearance of free, value-neutral, wonderful experimentation persists: plug in and play, copy and paste, upload and post. We have iPhones that record high-definition video (and software that allows us to edit it into slick movies), and online services like YouTube that allow us to show the world the fruits of our imagination. But the experimentation that is encouraged actually operates on these shallow planes. On deeper, more fundamental levels, it is strictly controlled.

Those controls have their roots in multiple, reinforcing causes. The growing popularity of what Jonathan Zittrain, in his book *The Future of the Internet*, called "tethered" devices is one of them: tethered, that is, because the devices are connected to their manufacturers long after they leave the showroom or store, and because they are built in such a way that no one but the manufacturers can change their internal workings. Because the impetus behind these tethered devices comes from users seeking protection from security threats, manufacturers seeking greater control over markets, and regulators looking to secure cyberspace as a whole, the momentum behind this direction is powerful and mutually reinforcing. Together, they risk gradually strangling the original conditions that nurtured innovation and the ethic of experimentation that gave rise to the Internet in the first place.

One of the more perverse examples of this dynamic is the way attempts to control cyberspace threatens security research, contributing to greater insecurities through the chilling effects associated with stringent copyright protections, such as those around the

Digital Millennium Copyright Act (DMCA) and its equivalents. The Electronic Frontier Foundation has documented numerous examples of security research being stifled, and of researchers veering away from contentious areas of investigation for fear of being held liable for breaching computer crime or copyright laws. In 2002, Secure Network Operations (SNOsoft) released a paper demonstrating security vulnerabilities in Hewlett-Packard's Tru64 Unix operating system. The company threatened SNOsoft with DMCA litigation. "After widespread press attention, HP ultimately withdrew the DMCA threat," noted the EFF. "Security researchers got the message, however – publish vulnerability research at your own risk." In 2003, publisher Wiley & Sons commissioned security researcher Andrew Huang to write a book about security flaws in Microsoft's xbox that he had discovered as a part of his Ph.D. research, but then dropped it because of liability concerns that the book could be treated as a "circumvention device," and thus in violation of the DMCA. The EFF has also written a critique on a draft Directive on Attacks Against Information Systems, a computer crime law currently being debated by the European Parliament, which the EFF says "threatens to create legal woes for researchers who expose security flaws." The EFF points to Article 3 of the draft directive, which makes it illegal to access information systems without authorization. In these and other cases like them, the legal risks around possible violations of intellectual property perversely stifle the research that is essential to securing the very foundation upon which those innovations rest.

Technology comes shrink-wrapped today, with stiff punishments for those caught trying to unwrap it. Nearly every software application, every tool downloaded, every app installed, and every DVD viewed is preceded by end-user licence agreements that list one prohibition after another. There is something profoundly disturbing about a culture in which in order to use technology individuals must first click "I agree" to such lengthy stipulations, and that restricts our communications behaviour, to say nothing of our native curiosity.

Never before have we had such a grand illusion of freedom through technology, when, in fact, that very freedom and technology are constrained by ever-expanding state laws and corporate regulations. This is not how it should be, or was meant to be. In an era when so much power is exercised beneath the surface of our technical systems, often deliberately hidden from scrutiny and shrouded in layers of deliberate obfuscation, a healthy curiosity about those systems is being actively discouraged. In a vital liberal democracy, citizens should be trained at an early age not only to use technology but also to understand it, to experiment with it, explore its hidden recesses, and shed light on those dimensions of the digital world where unchecked and unaccountable power resides. If by "hacking" we mean a healthy curiosity about technology, we need more, not fewer, hackers. Indeed, if experimentation around the technology of cyberspace were encouraged, not only would there be fewer unexposed vulnerabilities that create insecurity, there would not be a need for a reactionary phenomenon like Anonymous in the first place.

15.

Towards Distributed Security and Stewardship in Cyberspace

The whole human memory can be, and probably in a short time will be, made accessible to every individual . . . It need not be concentrated in any one single place. It need not be vulnerable as a human head or a human heart is vulnerable. It can be reproduced exactly and fully, in Peru, China, Iceland, Central Africa, or wherever else seems to afford an insurance against danger and interruption. It can have at once, the concentration of a craniate animal and the diffused vitality of an amoeba.

— H.G. Wells, "World Brain: The Idea of a Permanent Encyclopedia," 1937

Over half of the world's 7 billion people now share a single complex information and communications system — cyberspace — that functions, and functions very well, despite no grand blueprint or central point of control. Born as an experimental research network in universities, what used to be called the "Internet" has mushroomed, more by accident than design, to become the information and communications operating system for planet Earth. A mixed common-pool resource that cuts across political jurisdictions and the public and private sectors, cyberspace has become, as Marshall McLuhan foresaw, "our central nervous system in a global embrace."

This unprecedented global network produces a remarkable stream of innovations and social goods. Deep wells of knowledge, translated into multiple languages, are now instantly accessible to people around the world. H.G. Wells's description of a world encyclopaedia, written less than eighty years ago, is no longer the stuff of science fiction. Geolocational coordinates down to the level of centimetres are now available in the palm of a hand; instantaneous information sharing – "crowd-sourced" among connected individuals – holds out the potential of revolutionizing everything from election monitoring to disaster relief to predicting disease outbreaks; historic documents can be instantly translated into multiple languages, dramatically expanding the global pool of knowledge. And yet, as sweet as the fruits of cyberspace are, there are some that are poisonous. Malicious software that pries open and exposes insecure computing systems is developing at a rate beyond the capacities of cyber security agencies even to count, let alone mitigate. Data breaches of governments, private sector companies, NGOs, and others are now an almost daily occurrence, and systems that control critical infrastructure – electrical grids, nuclear power plants, water treatment facilities – have been demonstrably compromised.

These unfortunate by-products of an open, dynamic network are exacerbated by increasing assertions of state power. Insecurity, competition, and mounting pressures to deal with breaches, malware, and the other dark sides of cyberspace are driving such government interventions. Internet censorship at the national level, once thought to be impossible, is now the global norm, and governments race to develop cyber security strategies, including offensive cyber warfare capabilities. The 2012 leaks that provided details on U.S. and Israeli computer network operations that sabotaged Iranian nuclear enrichment facilities took few by surprise, as many suspected their hands in the Stuxnet virus in the first place. What was surprising was the calculated admission itself, the first instance of a government acknowledging – or at least not denying responsibility – an attack on critical infrastructure through cyberspace. Indeed, Stuxnet did cross the Rubicon.

Other countries are seeking advantage from the cyber criminal underground, stirring a hornet's nest of data theft and espionage from which they derive strategic intelligence and security benefits. Added to this dangerous brew is a mushrooming commercial market for offensive cyber attack capabilities. The global cyber arms trade now includes malicious viruses, zero-day exploits, and massive botnets. An arms race in cyberspace has been unleashed, and for every U.S. Cyber Command there is now a Syrian or Iranian cyber army equivalent. For every "Internet Freedom in a Suitcase," there is a justification put forward for greater cyberspace regulations and controls. We find ourselves in a situation where there are enormous profits to be made in developing capabilities to *deny* access to knowledge, prevent networks from functioning, or subvert them entirely. Fibre-optic surveillance and cyberspace disruption is now big business.

H.G. Wells was only half right: we have indeed created a kind of "world brain" – the problem is that it is an aggressive, insecure, and all-too-human one, and increasingly less the beautiful thing he imagined.

• • •

Faced with mounting problems and pressures to *do something*, too many policy-makers are tempted by extreme solutions. The Internet's de facto distributed regime of governance – largely informal and driven up to 2000 by decisions made by mostly like-minded engineers – has come under massive stress as a function of the Internet's rapid growth and insecurity. Proposals being debated in liberal democratic countries now include censoring the Internet in response to copyright violations; giving secretive signals intelligence agencies responsibility for securing cyberspace; loosening or eliminating judicial oversight around data sharing with law enforcement; and delegating Internet policing to the private sector. All are illustrations of a movement towards clamp down. These policies are antithetical to the principles of liberal democratic government and to the system of checks and balances and public accountability upon which it rests, and yet they are being put in place. They also legitimize the growing desire of autocratic and authoritarian regimes to subject cyberspace to territorialized controls, and the censorship and surveillance practices that go along with them. By our actions in the West, we contribute to this trend abroad. We preach about the need for closed autocratic societies to "open up," or, as Ronald Reagan famously thundered, to "tear down this wall," and yet vis-à-vis cyberspace we are contributing to state censorship and surveillance. Although states were once thought to be powerless in the face of the Internet, the giants have awoken from their slumber.

Left unchecked, these trends will result in the gradual disintegration of what is in the long-term interest of all citizens: an open and secure commons of information on a planetary scale. We stand at a crossroads, and there are several paths we can travel down. Fifty years from now, future historians may look back and say, "You know, there was that brief window in the 1990s and 2000s, when citizens came close to building that planetary library and global public sphere, and then let it slip from their grasp." The social forces leading us down the path of control and surveillance are formidable, even sometimes appear to be inevitable. But nothing is ever inevitable. The future has yet to be written. We face other extraordinary challenges, like climate change and global environmental degradation, issues that also appear at times to be so large and intractable, so requiring fundamental change (from us) as to be hopeless. In fact, the two spheres are intimately connected: we live in an increasingly compressed and interconnected political space on planet Earth, and to solve these problems, it is imperative that we have an open, shared, and equally accessible medium of global communications. We need more than ever to encourage, rather than stifle, the free flow of knowledge and the exchange of ideas, and cyberspace has provided us with that opportunity.

To protect planet Earth, we need to protect the Net.

As with environmental challenges, the solutions to problems vexing cyberspace are going to require approaches at multiple levels – local, national, and global. The articulation of an alternative vision of security, one that doesn't throw the baby out with the bath water, one that protects and preserves cyberspace as a dynamic and open and yet secure ecosystem, is urgently required. At the heart of this vision must be the elaboration of the proper rights, roles, and responsibilities for all who share in and sustain cyberspace, and it means ensuring that those rights, roles, and responsibilities are implemented and enforced. It is important to recall that cyberspace belongs to everybody and nobody in particular, that it is what we make of it, and that it requires constant tending.

• • •

Surely, one thinks, the challenges of an unprecedented planetary network of communications, of something so complex as global cyberspace, require a special cyber *theory* of some sort, something that rises to the scope and scale of this all-encompassing domain? Maybe. But maybe not. Instead, perhaps what is required is simply the application of some timeless principles and traditions.

There is an instinctive tendency in security-related discussions to default to *realpolitik* or Realism (and the theory that world politics is driven by competitive self-interest) with its state-centrism, top-down hierarchical controls, and the erecting of defensive perimeters to outside threats. In the creation of cyber commands, in spiralling arms races among governments, in "kill switches" on national Internets, and in the rising influence of the world's most secretive agencies into positions of authority over cyberspace, we see this tradition at play. As compelling as it may be, however, Realism and its institutional manifestations fit awkwardly in a world where divisions between inside and outside are blurred, where threats can emerge as easily from within as without, and where that which requires securing – cyberspace – is, ideally, a globally networked commons of information almost entirely in the hands of its users.

What is needed is an alternative cyber security strategy rooted in liberal democratic principles that takes account of the growing need for civic networks to share knowledge and to communicate. For many who would characterize themselves as part of global civil society, "security" is seen as anathema. In today's world of exaggerated threats and self-serving hyperbole from the computer security industry, it is easy to dismiss security as something to be resisted, rather than engaged. Securitization is generally associated with the defence industry, Pentagon strategists, and so forth, and many question whether employing the language of security only plays into the cyber-security military-industrial complex and the exercise of control. But the vulnerabilities of cyberspace are real, the underbelly of cyber crime undeniably huge and growing, an arms race in cyberspace escalating, and major governments are poised to set the rules of the road that may impose top-down solutions that subvert the domain as we know it. Dismissing these concerns as manufactured myths propagated by power elites will only marginalize civic networks from the conversations where policies are being forged.

Instead, civic networks need to be at the forefront of security solutions that preserve cyberspace as an open commons of information, and that protect privacy and support freedom of speech, while at the same time addressing the growing vulnerabilities that have produced a massive explosion in cyber crime. Can security and openness be reconciled? Aren't the two contradictory?

• • •

Not at all.

One alternative approach towards security that meshes with the core values and decentralized architecture of an open and secure cyberspace, and that has a long pedigree in political philosophy, is the "distributed" approach. It has roots in liberal political orders reaching back to ancient Greece and the Roman republic, and the late-medieval, early-Renaissance trade-based systems exemplified by the Venetians, Dutch, and English. But the fullest expression of distributed security is to be found in the early United States of America and the writings of the political philosophers who inspired the nation's founders, Montesquieu, Publius, and others. Although multi-faceted and complex, distributed security starts with building structures that rein in and tie down political power, both domestically and internationally, as a way to secure rights and freedoms. It puts forward what Johns Hopkins University Professor Daniel Deudney, author of *Bounding Power*, calls "negarchy" as a structural alternative to the twin evils of hierarchy and anarchy. In short, distributed security is the *negation* of unchecked and concentrated power, and, on the other side, recklessness and chaos.

At the core of this model are three key principles: mixture, division, and restraint. Mixture refers to the intentional combination of multiple actors with governance roles and responsibilities in a shared space; division to a design principle wherein no single actor is able to control the space in question without the co-operation and consent of others. As an approach to global cyberspace security and governance, each of these can provide a more robust foundation for the empty euphemism of "multi-stakeholderism," and principles upon which to counter growing calls for a single global governing body for cyberspace. Citizens, the private sector, and governments all have important roles to play in securing and governing cyberspace, but none to the exclusion or pre-eminence of the others.

Civic networks need to be players in the governance forums where cyberspace rules of the road are implemented. This is not an easy task. There is no single forum of cyberspace governance; instead, governance is diffuse and distributed across multiple forums, meetings, and standard-setting bodies at local, national, regional, and global levels. The acceptance of civil society participation in these rule-making forums varies widely, and the very idea is alien to some. Governments and the private sector have more resources at their disposal than citizens to attend these meetings and influence their outcomes. Civic networks will need to collaborate to monitor all of these centres of governance, open the doors to participation in those venues that are now closed shops, and make sure that "multi-stakeholder participation" is not just something paid lip service to by politicians, but something meaningfully exercised as part of a deliberate architecture.

The principle of restraint, however, is perhaps the most important and arguably the most threatened by overreaction. Securing cyberspace requires a reinforcement, rather than a relaxation, of restraint on power, including checks and balances on governments, law enforcement, intelligence agencies, and on the private sector. In an environment of big data, in which so much personal information is entrusted to third parties, oversight mechanisms on government agencies and involved corporations are essential.

Principles of restraint – sometimes called "mutual restraint" – can also help inform international cyberspace governance discussions concerning confidence- and security-building measures among states. Danger in cyberspace is real but to avoid overreaction, transparent checks and balances are required. Here, the link in the distributed security model between domestic and international processes is exceptionally clear. The more transparent the checks placed on concentrated power at the domestic level, the more adversaries abroad will have confidence in each other's intentions.

Distributed security also describes the most efficient and widely respected approach to security in computer science and engineering circles. It is important to remind ourselves that in spite of the threats, cyberspace runs well and largely without persistent disruption. On a technical level, this efficiency is founded on open and distributed networks of local engineers who share information as peers in a community of practice rooted in the university system (itself, a product of the liberal philosophy upon which distributed security rests). These folks need to be central during discussions about cyberspace governance.

The Internet functions precisely because of the absence of centralized control, because of thousands of loosely coordinated monitoring mechanisms. While these decentralized mechanisms are not perfect and can occasionally fail, they form the basis of a coherent distributed security strategy. Bottom–up, "grassroots" solutions to the Internet's security problems are consistent with principles of openness, avoid heavy-handedness, and provide checks and balances against the concentration of power. Part of a distributed security strategy should facilitate cooperation among largely scattered security networks, while making their actions more transparent and accountable. Rather than abolish this system for a more top-down approach, we should find ways to buttress and amplify it. Loosely structured but deeply entrenched networks of engineers, working on the basis of credible knowledge and reputation, whose mission and raison d'être is to focus on cyberspace and its secure functioning to the exclusion of all else, are essential to its longevity and security. We need to build out and provide space for those networks to thrive internationally rather than co-opt their talents for national security projects that create divisions and rivalry.

Part of a distributed security strategy must also include a serious engagement with law enforcement. These agencies are often stigmatized as the Orwellian bogeymen of Internet freedom (and in places like Belarus, Uzbekistan and Burma, they are), but the reality in the liberal democratic world is more complex. Many law enforcement agencies are overwhelmed with cyber crime, are understaffed and lack proper equipment and training, and have no incentives or structures to co-operate across borders. Instead of dealing with these shortcomings head on, politicians are opting for new "Patriot Act" powers that dilute civil liberties, place burdens on the private sector, and conjure up fears of a surveillance society. What law enforcement needs is not new powers, but new resources, capabilities, proper training, and equipment. Alongside those new resources, of course, the highest possible standards of judicial oversight and public accountability must be enforced.

The same basic premise of oversight and accountability must extend also to the private sector. Civic networks like those that helped spawn the Arab Spring are inherently transnational and have a vital role to play in monitoring the globe-spanning corporations that own and operate cyberspace. Persistent public pressure, backed by credible evidence-based research and campaigns – like the Electronic Frontier Foundation's privacy scorecard – are the best means to ensure the private sector complies with protection of privacy laws and human rights standards worldwide. Civic networks must also make the case that government pressures to police the Internet impose costly burdens on businesses and should be legislated only with the greatest reservations and proper oversight. The securitization of cyberspace may be inevitable, but what forms it takes is not.

• • •

If we are to continue to benefit from the common pooled resources that make cyberspace what it is — a planetary ecosystem in which no one central agency is in control — then all members of that ecosystem need to approach its maintenance in a deliberate and principled fashion. Here is where another tried and true approach might have broad utility for cyberspace: stewardship.

Cyberspace is less a pure public commons and more a mixed-pooled resource, with constantly emergent shared properties that benefit all who contribute to it. Does stewardship — generally defined as an ethic of responsible behaviour in regard to shared resources — have any relevance to such a domain? The first custodians of the Internet believed that it did. Even if they did not use the language of stewardship, the engineers and scientists who built and designed the Internet saw their roles very much as custodians of some larger public good.

In discussing the stewardship of cyberspace, one must remember that it is an entirely artificial environment; that is, without humans, cyberspace would not exist. This places us all in the position of joint custodianship: we can either degrade, even destroy cyberspace, or preserve and extend it. The responsibility is intergenerational, extending to those digital natives yet to assume positions of responsibility, but also linked to those who first imagined the possibilities for what something like cyberspace could represent. Imagine if H.G. Wells were here today to see how close we are to accomplishing his vision of a world encyclopaedia, only to see it carved up by censorship, surveillance, and militarization?

Governments, NGOs, armed forces, law enforcement and intelligence agencies, private sector companies, programmers, technologists, and average users must all play vital and interdependent roles as stewards of cyberspace. Concentrating governance of cyberspace in a single global body, whether at the UN or elsewhere, makes no sense. The only type of security that functions in an open, decentralized network is distributed security.

Stewardship happens constantly in cyberspace, even if not described as such. When Twitter unveiled a new national tweet removal policy, it felt obligated to justify its actions in terms of larger consequences, and the larger Internet community judged it accordingly. When companies like Google post transparency reports, listing government requests on user data or notices to remove content from cyberspace, these are acts of stewardship. As people entrust more and more data to third parties, how that information is handled, with whom it is shared, and what is communicated about how that data is treated, must be based on more than corporate self-interest and market considerations. Likewise, profiting from products and services that violate human rights, or that exacerbate malicious acts in cyberspace, is unjustifiable in a context of shared information and communication resources, regardless of how profitable such products and services might be. Justifying these sales on their being in compliance with local laws, as some companies have done, is a hollow and self-serving rationalization that fails the stewardship test of maintaining a global resource.

Generalized across the world, stewardship would moderate the dangerously escalating exercise of state power in cyberspace by defining limits and setting thresholds of accountability and mutual restraint. The alarming trend of even liberal democratic governments engaging in mass surveillance without judicial oversight contradicts the very essence of cyberspace as an open global commons. Governments have an obligation to establish the playing field, ensure that malicious acts are not tolerated within their jurisdictions, and set the highest possible standards of self-restraint vis-à-vis censorship and surveillance. Privacy commissioners and other regulatory and competition oversight bodies are critical to stewardship in cyberspace, as more and more information and responsibilities are delegated to the private sector. In an era when "national security" is so often used to justify extraordinary intrusions on individual privacy, checks and balances are essential.

Universities have a special role to play as stewards of an open and secure cyberspace as it was from "the University" that the Internet was born, and from which its guiding principles of peer review and transparency were founded. Protected by academic freedom, equipped with advanced research resources that span the social and natural sciences, and distributed across the planet, university-based research networks could be the ultimate custodians of cyberspace.

Finally, stewardship in this realm requires an attitudinal shift among users as to how they approach cyberspace. For most of us, it is William Gibson's "consensual hallucination" – always on, always working, 24/7, like running water. This attitude shift will not be easy. There are considerable disincentives for average people to "lift the lid" on the technology. While we are given extraordinary powers of creativity with cyberspace, walled gardens restrict what we can actually do with it. Busting down these walls has to be at the heart of every citizen's approach to cyberspace. We don't all need to learn computer code, but we do need to move beyond sending emails or tweets out into the ether without understanding with whom, beyond the immediate recipient, they are shared and under what circumstances.

We are at a crossroads. Mounting cyber threats and an escalating arms race are compelling politicians to take urgent action. In the face of these concerns, those who care about liberal democracy on a global scale must begin to articulate a compelling counter-narrative to reflexive state and corporate control over cyberspace. To be sure, distributed security and stewardship are not panaceas. They will not cease the exercise of power and competitive advantage in cyberspace. They will not bring malicious networks to their knees, or prevent cutthroat entrepreneurs from exploiting the domain. But, as a vision of ethical behaviour in cyberspace, they will raise the bar, set standards, and challenge the players to justify their acts in more than self-interested terms. Above all, they will focus collective attention on how best to sustain a common communications environment on a planetary scale in an increasingly compressed political space.

Decisions made today could take us down a path where cyberspace continues to evolve into a global commons that empowers individuals through access to information and freedom of speech and association, or they could take us in the opposite direction. Developing models of cyber security that deal with the dark side, while preserving our highest aspirations as citizens, is our most urgent imperative.

NOT AN EPILOGUE

People often ask me what the inspiration was for the Citizen Lab. Admittedly, doing what we do — a kind of *X-Files* meets academia — is highly unusual. But it has been no accident.

Although there have been many formative experiences along the way, one of the most important was an opportunity I had as a graduate student in the 1990s, when I was seconded to the Canadian Ministry of Foreign Affairs as a consultant for an obscure agency called the Verification Research Unit (VRU) headed by a retired Canadian Air Force colonel, Ron Cleminson. Run like a private fiefdom by the iconoclastic veteran, the VRU engaged in groundbreaking studies on arms control, particularly the often troubling question of how to verify whether parties to an arms control agreement were playing by the rules or cheating. Interested in technology and international security as a graduate student, I was contracted by Cleminson to explore how the then emerging commercial market for satellite reconnaissance technology could assist in the verification of arms control agreements.

My VRU experience suggested the potential of revolutionary changes in information and communications technologies to have a major impact on international security. New satellites were being launched by the governments of France, Canada, and other states that only a few years prior would have been the most guarded secrets of the intelligence community, but now imagery from them was being shown to the general public and offered for sale.

The implications of all of this hit me shortly after the Gulf War in the early 1990s. Taken aside by a member of the VRU to a locked, windowless room, I was shown highly sophisticated spy-satellite imagery of a couple of scared Iraqis frantically burying drums in the desert. Laid on the desk before me were high-resolution images taken from a KH-11 U.S. spy satellite, orbiting the earth in synchronicity with the path of the sun so that the surface illumination was nearly the same in every picture. Familiar today to viewers of movies like the *Bourne* series, the imagery was astoundingly sharp – a ground resolution of six centimetres – so sharp that I could clearly make out the expressions on the Iraqis' faces. At the time, these images were highly classified, and I did not have clearance to see them.

Looking today at my iPhone's RunKeeper app, which tracks my jogging route down to the level of metres in real time, that moment in the VRU office seems so quaint. How soon, I wondered, given current technological trajectories, would KH-11 imagery be available to the entire world? How long could it remain in the shadows?

While at the VRU I attended meetings, workshops, and conferences that involved fascinating applied policy work, much of it highly interdisciplinary. Nuclear, chemical, and biological engineers worked alongside policy analysts and lawyers; government officials, private sector representatives, and people from academia, all with vast but very different experiences, collaborated on international security projects. In the mid-1990s – the World Wide Web barely off the ground and cyber security on pretty much no one's mind – I attended a conference organized by Cleminson with the prescient title "Space and Cyberspace: Prospects for Arms Control." In attendance, an extraordinary cast: an analyst who had handed John F. Kennedy the overhead imagery from the Cuban missile crisis in 1962; a scientist working at Sandia National Labs tracking down the Aum Shinrikyo cult, the Japanese terrorist group that had dumped

Sarin nerve gas in the Tokyo subway and who some suspected had purchased property in Australia to test a primitive nuclear device; a technician working on Canada's RADARSAT satellite, whose synthetic aperture radar imaging could peer through clouds and darkness from space to resolve objects on the surface of the earth.

A major inspiration that would later inform the Citizen Lab's "mixed methods" approach came via my experiences researching the technical work around the Comprehensive Test Ban Treaty (CTBT) negotiations, which at that time were occurring through the venue of the United Nations Conference on Disarmament. The process involved nuclear, radiological, chemical, seismic, and imagery specialists from about a dozen countries whose mission was to provide a blueprint for a planetwide surveillance network to verify compliance to a possible CTBT, then under negotiation. The process was highly politicized – with the United States and its allies continuously trying to stall negotiations, in my view – and by the time I dropped into the process, the scientists had been meeting for years, knew each other as close friends. Their plans for total Earth surveillance were so airtight that, as one participant joked, "if an ant farted anywhere on earth, we'd know about it." The architecture for the CTBT verification system included a worldwide network of seismic sensors; radionuclide sniffing stations that would suck up the air and detect the slightest wisp of anything nuclear; space-based radar, optical, and infrared satellites; and even underwater hydro-acoustic sensors, to capture nuclear tests that might be conducted in the ocean's depths. Though CTBT has never received enough state ratifications to enter into force, the image of a worldwide network of sensors combining various technological platforms, from undersea to outer space, all meant to check and constrain cheating around nuclear testing and build confidence and security for the planet, stuck with me deeply and still influences how I think global cyber security should be implemented.

When the Citizen Lab was founded in 2001, I had in mind a similar image, a planetary network with data collected by researchers and field investigators, this time all related to cyberspace openness and security. When we started we were the only game in town, but over time we built up a network of collaborations with individuals and other university research centres that continues to grow.

Now, more than ten years later, the situation has changed substantially. I have just received an invitation from Harvard's Jonathan Zittrain (also one of the founders of OpenNet Initiative) to attend a preliminary planning meeting for something he is calling the Internet Health Organization (IHO). His vision for the IHO is similar to my own: a distributed network of research centres monitoring the health of the Internet using a variety of methods and approaches. Included in the preliminary meeting are numerous groups who have undertaken highly imaginative and constructive projects in this broad area: Herdict, a project that collects and disseminates real-time, crowd-sourced information about Internet filtering, denial of service attacks, and other blockages; M-Lab, an "open, distributed server platform for researchers to deploy Internet measurement tools"; and StopBadware, which "aims to make the Web safer through the prevention, mitigation, and remediation of badware websites," among others.

Just after Zittrain's invitation came another, this time from the European Commission, which was planning a meeting to discuss the development of a "European Capability for Situational Awareness" platform. According to their invitation, the aim is to gather "reliable and real-time or almost real-time information concerning human rights violations and/or restrictions of fundamental freedoms in connection with the digital environment," and to determine "what is happening in the Net, in terms of network connectivity and traffic alterations or restrictions."

Projects like these, and numerous others sprouting up around the globe, show that the mission of the Citizen Lab is resonating with others, and that we are not alone. Will these collective efforts have an impact? Will they be enough to ensure cyberspace remains an open and secure commons of information that helps citizens reach their highest aspirations in this increasingly interconnected and constrained political space?

Just as I am about to send my manuscript to the publisher, a major news story breaks: Syria pulls the plug on the Internet. An announcement on Syrian state TV says that "maintenance technicians are working to fix the problems," but many suspect the drastic measure is a prelude to a major armed assault on the opposition. The Syrian Internet shutdown comes only a few days before a major meeting in Dubai of the International Telecommunication Union, which has stoked fears about the growing role of states and the UN in Internet governance. The two are not unrelated: the forces moving us towards enclosure, secrecy, and an increasingly dangerous arms race are powerful and grow daily. Sometimes it seems futile to resist them.

In hindsight, the organizers of that May 2012 Calgary conference may have been on to something with their title, "Nobody Knows Anything." We do know an awful lot these days, with data exploding all around us and information at our fingertips as never before. But the fact remains that nobody really knows where the dark forces in cyberspace are driving us, and whether they can be tamed. We can only keep probing beneath the surface, lifting the lid, and trying to get a handle on this domain that we have created, remembering that cyberspace is, after all, what we together make of it.

ONE HUNDRED AND FOUR YEARS OF ANGLO-AMERICAN SURVEILLANCE:

A Selected Timeline

1909: U.K. Secret Service Bureau is founded

The Secret Service Bureau starts out as a joint initiative of the British Admiralty and the War Office, with the navy focusing on foreign intelligence and the army on counter-espionage. With the outbreak of World War I in 1914, the division is formalized into two new agencies: MI6 and MI5 respectively. In 1919, the U.K. Cabinet Secret Services Committee recommends consolidating all of the signals intelligence activities into a single peacetime codebreaking agency called the Government Code and Cypher School (GC&CS). GC&CS is renamed the Government Communications Headquarters (GCHQ) in **June 1946**.

1943: BRUSA Agreement is signed

The British and American governments agree to have the U.S. War Department and the British Government Code and Cypher School (GC&CS – later, the Government Communications Headquarters, GCHQ) share signals intelligence. The agreement is followed up in **March 1946** with the "United Kingdom-United States of America Agreement" (UKUSA), which is later extended to include the signals intelligence agencies of Canada, Australia, and New Zealand – collectively known as the "Five Eyes." The Five Eyes working arrangement continues up to the present day.

April 25–June 26, 1945: Interception at the San Francisco Conference

Formally known as the United Nations Conference on International Organization, the San Francisco Conference is the international meeting that establishes the United Nations. According to intelligence historian James Bamford, the U.S. offered to host the meetings to eavesdrop more efficiently on visiting foreign officials. Cable traffic to and from foreign capitals is passed on to U.S. codebreakers and analysts by the Western Union and other commercial telegraph carriers under wartime censorship laws still in effect at the time.

1946: Communications Security Establishment of Canada (CSEC) is founded

CSEC is originally founded as the Communications Branch of the National Research Council, which has its origins in World War II–era Canadian signals intelligence activities. The highly secretive agency remains largely unknown to the Canadian public until a television documentary exposes its operations in 1974. The Canadian government only officially acknowledges CSEC's existence on **September 22, 1983**.

November 4, 1952: The U.S. National Security Agency (NSA) is created

Following unhappiness about poor signals intelligence analysis during the Korean War, the Truman Administration creates a new secret agency, the NSA, in November 1952 with the dual responsibility of domestic government communications security and foreign communications intelligence collection, analysis, and dissemination. Known informally by employees as "No Such Agency" the very existence of the NSA is not publicly acknowledged or discussed by government officials through most of the Cold War.

1954: RAND's Project Feedback is proposed

In 1954, the RAND project (the precursor to today's RAND corporation) formally proposes Project Feedback, an unmanned satellite operating 300 miles above the Earth that could photograph the Earth's surface at a resolution of 100 feet and transmit the images through a television camera. The proposal develops into the "Advance Reconnaissance System – Weapon System 117 (WS-117)," which is contracted out to Lockheed Martin, leading eventually to the Corona series of satellites.

June 1959: The U.S. Corona program is launched

The Corona program is a series of U.S. optical-imaging reconnaissance satellites first launched in June 1959 and continued until 1972. The Corona satellites are designed to take photographs of the Earth from space and then eject capsules of film that are scooped up by airplanes upon returning to the Earth's atmosphere. In **1976**, the National Reconnaissance Office begins using the KH-11 Kennan reconnaissance satellites, later renamed Crystal, which employed electro-optical digital imaging and were able to provide real-time transmission of images. The latest generation of optical imaging satellites provide a ground resolution of roughly 6 centimetres.

1968: U.S. "Canyon" satellites are launched

The Canyon satellites are known for being the first U.S. satellite system for the purposes of experimental communications intelligence. The satellites were able to pick up voice and data traffic from the Earth's orbit using large parabolic reflecting dishes that are unfolded once in orbit. During the same period, the NSA experiments with a series of ground station receivers that pick up microwave signals as they bounce off the Moon and ricochet back to Earth.

1970: U.S. Rhyolite (later named Aquacade) is launched
The first Rhyolite satellite is launched in 1970 in geosynchronous orbit and sends data to a remote location in Australia (Alice Springs, later renamed Pine Gap) to avoid Soviet detection. The Rhyolite satellite surveillance program is a breakthrough in signals intelligence quality and quantity, vastly superior to ground-based eavesdropping and allowing unfettered access to tens of thousands of telephone calls, data transmissions, and telemetry signals.

June 19, 1972: U.S. vs United States District Court
In a landmark Supreme Court ruling delivered by Mr. Justice Powell, the court rules against the Nixon administration, decides warrants are required for domestic surveillance and therefore upholds the Fourth Amendment. According to the majority opinion, "The danger to political dissent is acute where the Government attempts to act under so vague a concept as the power to protect 'domestic security.' Given the difficulty of defining the domestic security interest, the danger of abuse in acting to protect that interest becomes apparent."

January 27, 1975: U.S. Senate Church Committee is established
Following the Watergate scandal, and several high-profile media revelations of apparent abuse and overreach by U.S. intelligence agencies, the Senate votes to create a special investigating body, the United States Senate Select Committee to Study Governmental Operations with Respect to Intelligence Activities (also known as the Church Committee after chairman, Senator Frank Church). The Church Committee investigates illegal intelligence gathering by the CIA, NSA, and FBI, uncovers unlawful domestic spying by the NSA, and recommends several important reforms. At the conclusion of the committee, Chairman Church remarks prophetically, "The NSA's capability at any time could be turned around on the American people, and no American would have any privacy left, such is the capability to monitor everything: telephone conversations, telegrams, it doesn't matter."

May 15, 1975: Project Shamrock is terminated
Established in August 1945, this secret surveillance program is the peacetime continuation of World War II censorship laws that made international message traffic available to the U.S. military. Project Shamrock collects all telegraphic communications with the help of participating international carriers, including RCA Global, ITT World Communications, and Western Union International. The companies provide the NSA and its predecessor, the Armed Forces Security Agency, with microfilm copies of all incoming, outgoing, and transiting telegraphs daily. In its last months,

the project collects upwards of 150,000 messages per month. The NSA produces reports based on this collection and submits them to domestic law enforcement and intelligence agencies, including the Department of Defense, the CIA, the FBI, the Secret Service, and the Bureau of Narcotics and Dangerous Drugs (BNDD). The project is terminated after the Church Committee's investigations. In its final report, the committee concludes that "SHAMROCK was probably the largest governmental interception program affecting Americans ever undertaken."

1978: The Foreign Intelligence Services Act (FISA) is signed into law

FISA is among the reforms of the Church Committee investigation. The Act, for the first time, lays out the procedures around military intelligence and electronic surveillance for foreign intelligence gathering. The Act prohibits the warrantless collection of raw telegrams as under Project Shamrock, as well as arbitrary watch lists of American citizens. The Act also establishes the U.S. Foreign Intelligence Surveillance Court. Under FISA, the NSA is required to obtain a secret warrant from the court in order to target an American citizen or permanent resident alien, by showing that the target is either an agent of a foreign government, or participating in espionage or terrorism.

July 11, 2001: European Parliament releases final Echelon report

Following the investigations of author Duncan Campbell into national and international telecommunications interception networks, the European Parliament sets up a special temporary committee into the "Echelon" system. Its final report, published in 2001, provides details on a global network of computers and a keyword searching program operated cooperatively by the national signals intelligence of the U.S., U.K., Canada, New Zealand, and Australia, which searches through millions of intercepted communications collected by ground, sea, air, and space-based systems.

September 11, 2001: The 9/11 terrorist attacks on the United States

The 9/11 attacks on the United States usher in a swift and dramatic reorientation of how security agencies operate in the United States and among allied countries. As ex-NSA analyst J. Kirk Wiebe recalls, "everything changed at the NSA after the attacks on September 11. The prior approach focused on complying with the ... FISA. The post-September 11 approach was that NSA could circumvent federal statutes and the Constitution as long as there was some visceral connection to looking for terrorists." Several new broad surveillance programs, including mass collection of U.S. domestic communications, are authorized by then President George W. Bush, known collectively as the "President's Surveillance Program." On **October 4, 2001**, Bush authorizes a covert warrantless domestic surveillance program in which telecommunication companies are compelled to turn over bulk user data to the NSA.

October 26, 2001: President George W. Bush signs into law the U.S. Patriot Act

The Patriot Act ushers in a series of sweeping legislative changes in the United States which significantly enhance the surveillance powers of law enforcement and intelligence agencies, and lower checks and balances around protection for civil liberties. The Patriot Act is passed with great haste and little public debate only weeks after the 9/11 attacks and provides a model for other countries to follow in passing their own anti-terrorism legislation. Among the Patriot Act's most important provisions is Section 215, which permits the government to obtain business records and/or "tangible things" with the approval of the FISA court as long as the information is sought for "an authorized investigation . . . to protect against international terrorism or clandestine intelligence activities." Section 215 comes with a gag order that prevents recipients of such requests from ever divulging to anyone at any time that they received them. Later, the NSA uses the Section 215 provisions to justify bulk collection of the phone records of all U.S. citizens.

December 18, 2001: Canada's Anti-terrorism Act is passed by the Liberal government

A few months after the Patriot Act, Canada passes Bill C-36, the Anti-Terrorism Act that gives law enforcement and intelligence agencies new powers, including the authorization of warrantless interception of foreign communications under an amendment to the National Defence Act and the relaxing of electronic surveillance requirements under the Criminal Code. Bill C-36 stipulates that CSEC can collect information from "the global information infrastructure" about the "capabilities, intentions or activities of a foreign individual, state, organization or terrorist group, as they relate to international affairs, defence or security." A second part of its mandate focuses on security of information infrastructures in Canada, while a third new set of responsibilities authorized by Bill C-36 specifies CSEC should assist federal law enforcement and security agencies "in performance of their lawful duties," and thus opens up possible surveillance of Canadian communications and domestic surveillance.

January 2002: The U.S. Defense Advanced Research Project Agency (DARPA) establishes the Information Awareness Office (IAO) and the Total Information Awareness (TIA) program

The specific aim of the TIA is to gather (without a warrant) the personal information of all U.S. citizens, including emails, credit card transactions, phone calls, medical records, and any other data, and then analyze it for suspicious activity associated with possible terrorism-related acts. Following controversy about the privacy implications of the program, Congress discontinues funding of the TIA in 2003. However, the program's main elements continue under different code names and are funded by "black budgets."

2005–2006: *New York Times* on Mark Klein's AT&T revelations, and *USA Today* on Telecoms

In **April 2006**, Mark Klein, a technician at AT&T's Internet Exchange Point in San Francisco, reveals that AT&T is operating an eavesdropping facility for a warrantless NSA surveillance program. The Klein revelations follow closely after a December 2005 *New York Times* story on the Bush warrantless surveillance program. On May 11, 2006, *USA Today* reveals that AT&T, Verizon, and Bell South all handed over U.S. citizen call records to the NSA. In response, the U.S. FISA Amendments Act was signed into law by President Bush in 2008. The new legislation transforms the FISA to allow warrantless surveillance for the purposes of acquiring foreign intelligence information. It also gives telecommunication companies retroactive immunity for cooperating with the Bush administration's warrantless wiretapping program. On December 30, 2012, President Obama signs the FISA Amendments Act Reauthorization Act of 2012, which extends the Amendments Act until 2017.

June 6, 2013: Snowden documents reveal details about the Verizon mass data collection program

In the first of a series of stunning disclosures based on documents leaked by former NSA contractor Edward Snowden, *The Guardian* reveals details about a U.S. Government order requiring Verizon to turn over millions of telephone records for a three-month period beginning in April 2013. The document confirms for the first time that the Obama administration had been covertly collecting U.S. communications records in bulk, without a warrant, and in spite of any suspicion or evidence of wrongdoing.

June 6, 2013: Snowden documents reveal PRISM

The Guardian and *Washington Post* report that the NSA has direct access to the servers of nine major U.S. internet companies: Microsoft, Google, Yahoo!, Facebook, PalTalk, YouTube, Skype, AOL, and Apple. This access is a part of a clandestine NSA-operated mass electronic surveillance data mining program called PRISM, which was launched in 2007, under the supervision of the FISA Court. PRISM allows authorities to collect a wide range of information, such as email, video and voice chat, photos, VoIP chats, file transfers, stored data, social networking details, and more. The participating companies initially deny knowledge of PRISM, but are also legally prevented from disclosing any knowledge as a consequence of gag orders.

June 10, 2013: The *Globe and Mail* reports that the Canadian Defence Minister authorized a "secret electronic eavesdropping program" in 2011

The *Globe and Mail* reports that the Canadian Defence Minister, Peter MacKay, had signed a ministerial directive which renewed the government's "metadata" surveillance program on **November 21, 2011**. According to the *Globe*, the "secret eavesdropping program . . . scours global telephone records and Internet data trails – including those of Canadians – for patterns of suspicious activity." The secret program is first approved in 2005 under the Liberal government, but is suspended by CSEC in 2008 after retired Supreme Court Judge Justice Charles Gonthier raises questions about the program. The *Globe* reports that Justice Gonthier's major concern was whether the program could result in warrantless surveillance of Canadians.

June 11, 2013: Snowden documents reveal "Boundless Informant"

Based on documents from Snowden, *The Guardian* reports on NSA's Boundless Informant program, which provides details and maps by country of the amount of information the NSA collects from computer and telephone networks. "The focus of the internal NSA tool is on counting and categorizing the records of communications, known as metadata, rather than the content of an email or instant message." The leaked documents reveal that the NSA had collected 97 billion pieces of intelligence from computer networks worldwide in March 2013, with the largest amount of intelligence being gathered from Iran, followed by Pakistan, Jordan, Egypt, and India.

July 21, 2013: Snowden documents reveal "Operation Tempora"

Documents leaked by Snowden reveal that U.K. spy agency, GCHQ, has access to a network of cables that carry a huge portion of the world's telephone and Internet communications, and is collecting and processing vast quantities of communications data in coordination with and compensated by the NSA. The operation, codenamed Tempora, was launched in August 2011. On August 2, 2013, the U.K. *Independent* reveals that Tempora's industry partners included BT, Verizon Business, Vodafone Cable, Global Crossing, Level 3, Viatel and Interoute. Tempora collects recordings of telephone calls, contents of emails, Facebook entries, and the history of a user's access to websites. The leaked documents show that by the summer of 2011, GCHQ has probes attached to more than 200 internet links, each carrying data at 10 gigabits a second and is working on expanding their processing capacity to 100 gigabits a second.

July 31, 2013: Snowden documents reveal "XKeyscore"

Snowden releases leaked documents to *The Guardian* revealing details about the NSA's XKeyscore program, a desktop system used to mine agency databases on an individual and obtain metadata and communications content of a user's Internet activity. *The Guardian* reports that the program could allow analysts to surveil someone over the Internet in "real time" – supporting Snowden's earlier assertion made to *The Guardian* that he "could wiretap anyone, from you or your accountant, to a federal judge or even the president, if I had a personal email."

August 21, 2013: CSEC Commissioner reports on CSEC's surveillance activities

In a rare public rebuke, the CSEC Commissioner releases his 2012–2013 annual report containing his findings that Canadians may have been unlawfully subject to CSEC's surveillance activities. The Commissioner writes: "I had no concern with respect to the majority of the CSEC activities reviewed. However, a small number of records suggested the possibility that some activities may have been directed at Canadians, contrary to law. A number of CSEC records relating to these activities were unclear or incomplete. After in-depth and lengthy review, I was unable to reach a definitive conclusion about compliance or non-compliance with the law."

August 30, 2013: Leaked Snowden documents shed light on "black budgets"

The *Washington Post* publishes a summary of leaked top secret documents that describe in detail the "black budget" breakdown of major intelligence operations involving the CIA, NSA, and other agencies. Among the disclosures are details on U.S. offensive operations in cyberspace. The documents show that the U.S. government engaged in 231 offensive cyber-operations in 2011, mostly undertaken by a unit within the NSA called Tailored Access Operations (TAO), and that the latter had managed to place 85,000 surreptitious "implants" into computers worldwide. The documents also show the NSA spent $25.1 million to purchase malicious software – "zero days" – from private companies.

September 6, 2013: Leaked Snowden documents reveal NSA, GCHQ plans to defeat crypto

The Guardian, *New York Times*, and *Pro Publica* publish leaked top secret documents that reveal the NSA and GCHQ worked extensively to defeat and systematically weaken encryption and other security protocols worldwide. Known by the codenames "Bullrun" in the United States and "Edgehill" in the United Kingdom, the efforts involved cooperation with tech companies to insert secret access points or "back doors" into products, and covert involvement in industry forums to shape outcomes and have weak standards adopted internationally.

NOTES

Portions of *Black Code* have been inspired by or drawn from previous publications, including "Contesting Cyberspace and the Coming Crisis of Authority" (with Rafal Rohozinski) in Ronald Deibert, John Palfrey, Rafal Rohozinski, and Jonathan Zittrain (eds.) *Access Controlled: The Shaping of Power, Rights and Rule in Cyberspace* (Cambridge: MIT Press, 2010); "Meet Koobface, Facebook's Evil Doppelgänger," (with Rohozinski), *Globe and Mail* (November 12, 2010); "Access Contested: Toward the Fourth Phase of Internet Controls," (with Palfrey, Rohozinski, and Zittrain), in *Access Contested: Security, Identity, and Resistance in Asian Cyberspace* (Cambridge: MIT Press, 2011); "Liberation vs Control: The Future of Cyberspace," (with Rohozinski), *Journal of Democracy*, 24, 1 (October 2010), pp. 43-57; "The Growing Dark Side of Cyberspace (. . . and What To Do About It)," *Penn State Journal of Law & International Affairs* (volume 1, no. 2, 2012).

PREFACE

1   **CSEC, Canada's version of the U.S. National Security Agency:** Communications Security Establishment Canada's (CSEC) mandate was updated under Canada's Anti-terrorism Act of December 2001. The Act stipulates that CSEC collect information from "the global information infrastructure" about the "capabilities, intentions, or activities of a foreign individual, state, organization, or terrorist group, as they relate to international affairs, defence, or security." A second part of its mandate focuses on security of information infrastructures in Canada, while a

third specifies CSEC should assist federal law enforcement and security agencies "in performance of their lawful duties." Details are in Anti-terrorism Act, SC 2001, c.41, s.102, codified as National Defence Act, RSC 1985, C.N-5, s, 273.61–273.7.

CSEC is Canada's partner in the so-called Five Eyes alliance of signals intelligence agencies that includes the United States (National Security Agency), the United Kingdom (Government Communications Headquarters), Australia (Defence Signals Directorate), and New Zealand (Government Communications Security Bureau). See Martin Rudner, "Canada's Communications Security Establishment, Signals Intelligence and Counter-terrorism," *Intelligence and National Security* (2007); James Bamford, *Body of Secrets: Anatomy of the Ultra-Secret National Security Agency* (New York: Anchor Books, 2002); and Jeremy Littlewood, "Accountability of the Canadian Security Intelligence Community Post 9/11: Still a Long and Winding Road?" in ed. Daniel Baldino, *Democratic Oversight of Intelligence Services* (Annandale, NSW: Federation Press, 2010).

2 **The Citizen Lab did not trespass or violate anything:** The ethical and legal issues underpinning the Citizen Lab's research are discussed in Masashi Crete-Nishihata and Ronald J. Deibert, "Blurred Boundaries: Probing the Ethics of Cyberspace Research," *Review of Policy Research* 28 (2011): 531–537.

3 **9/11 ripped into all of that and left us all reeling:** See Ronald J. Deibert, "Black Code: Censorship, Surveillance, and Militarization of Cyberspace," *Millennium: Journal of International Studies* 32, no. 3 (2003).

4 **in a *Globe and Mail* op-ed:** Ronald Deibert, "The Internet: Collateral Damage?", *Globe and Mail*, January 1, 2003, http://www.theglobeand-mail.com/commentary/the-internet-collateral-damage/article790542/.

5 **Another word, a few words actually, about the title:** Lawrence Lessig's, *Code and Other Laws of Cyberspace* (New York: Basic Books, 1999). Key McLuhan works are *The Gutenberg Galaxy: The Making of Typographic Man* (Toronto: University of Toronto Press, 1962) and *Understanding Media: The Extensions of Man* (New York: McGraw-Hill, 1964). Those of Harold A. Innis include *Empire and Communications* (Toronto: University of Toronto Press, 1950) and *Bias of Communications* (Toronto: University of Toronto Press, 1951). My take on Innis can be found in Ronald J. Deibert,

"Harold Innis and the Empire of Speed," *Review of International Studies* 25, no. 2 (1999). I wrote about media ecology theory and world order transformation in my first book, *Parchment, Printing and Hypermedia: Modes of Communication in World Order Transformation* (New York: Columbia University Press, 1997). Joshua Meyrowitz uses the metaphor of media as "environments" in *No Sense of Place: The Impact of Electronic Media on Social Behavior* (New York: Oxford University Press, 1985).

7   **The science fiction writer Arthur C. Clarke argued:** Clarke's comments about technology are part of his "three laws" of prediction and are outlined in Arthur C. Clarke, *Profiles of the Future: An Inquiry Into the Limits of the Possible* (London: Gollancz, 1962).

INTRODUCTION:
CYBERSPACE: FREE, RESTRICTED, UNAVOIDABLE

11   **Connectivity in Africa:** Information on Internet connectivity and growth rates is collected at Internet World Stats: Usage and Population Statistics, http://www.internetworldstats.com/stats.htm.

14   **Few of us realize that data stored by Google . . . are subject to the U.S. Patriot Act:** The official title of the Patriot Act is "Uniting and Strengthening America by Providing Appropriate Tools Required to Intercept and Obstruct Terrorism (USA PATRIOT) Act of 2001." The full Act can be found at http://www.gpo.gov/fdsys/pkg/PLAW-107publ56/pdf/PLAW-107publ56.pdf. See also "USA Patriot Act," Electronic Privacy Information Center, http://epic.org/privacy/terrorism/usapatriot/default.html.

14   **Mobile devices are what Harvard's Jonathan Zittrain:** Jonathan Zittrain warns about the shift towards "tethered appliances" in *The Future of the Internet and How to Stop It* (New Haven: Yale University Press, 2008).

16   **Botnets . . . can be rented from public forums and websites:** A price list of illicit products and services sold in the Russian cybercrime underground is documented in this Trend Micro report: Max Goncharov, "Russian Underground 101," *Trend Micro*, 2012, http://www.trendmicro.com/cloud-content/us/pdfs/security-intelligence/white-papers/wp-russian-underground-101.pdf. Many more details about cyber crime are provided in Chapter 8.

18   **The OpenNet Initiative (ONI) ... notes that roughly 1 billion Internet users:** The ONI was founded in 2002 as a partnership between the Citizen Lab at the Munk School of Global Affairs, University of Toronto, Berkman Center for Internet & Society at Harvard University, and the Advanced Network Research Group at the University of Cambridge, U.K. (later, the SecDev Group) by myself, Rafal Rohozinski, John Palfrey, and Jonathan Zittrain. The ONI's publications can be located at http://opennet.net/. The ONI estimates that in 2012, more than 620 million people lived in censored jurisdictions; see "Global Internet Filtering in 2012 at a Glance," OpenNet Initiative, April 3, 2012, http://opennet.net/blog/2012/04/global-internet-filtering-2012-glance.

     ONI has documented the use of Western-made software for Internet filtering in the Middle East and North Africa in Helmi Noman and Jillian C. York, "West Censoring East: The Use of Western Technologies by Middle East Censors, 2010–2011," http://opennet.net/west-censoring-east-the-use-western-technologies-middle-east-censors-2010-2011.

19   **Dissidents in the United Arab Emirates and Bahrain:** Instances of U.A.E. and Bahrain dissidents being targeted by British- and Italian-produced network intrusion kits have been reported in Vernon Silver, "Spyware Leaves Trail to Beaten Activist Through Microsoft Flaw," Bloomberg News, October 10, 2012, http://www.bloomberg.com/news/2012-10-10/spyware-leaves-trail-to-beaten-activist-through-microsoft-flaw.html; and Vernon Silver, "FinFisher Spyware Reach Found on Five Continents: Report," Bloomberg News, August 8, 2012, http://www.bloomberg.com/news/2012-08-08/finfisher-spyware-reach-found-on-five-continents-report.html. More details about this emerging marketplace can be found in Chapter 13.

1: CHASING SHADOWS

21   **So began the story of GhostNet:** Both the GhostNet and Shadows investigations were done under the auspices of the Information Warfare Monitor Project (2002–2011), a collaboration between the Citizen Lab at the Munk School of Global Affairs, University of Toronto and the Advanced Network Research Group at University of Cambridge, U.K. (later, the SecDev Group). Rafal Rohozinski was a co-principal investigator on the Information Warfare Monitor and one of the co-authors on both reports. Portions of the GhostNet field/technical research were carried out by Dr. Shishir Nagaraja of Cambridge University. Nagaraja and his

supervisor, Dr. Ross Anderson, released their own report coinciding with our GhostNet publication: Ross Anderson and Shishir Nagaraja, *The Snooping Dragon: Social-malware Surveillance of the Tibetan Movement*, Cambridge University Computer Laboratory Technical Report, 2009. The Shadowserver Foundation collaborated with the Information Warfare Monitor on the Shadows report, and Steven Adair was a co-author on that report. We documented our GhostNet and Shadows investigations in Information Warfare Monitor, Tracking GhostNet: Investigating a Cyber Espionage Network, March 29, 2009, http://www.scribd.com/ doc/13731776/Tracking-GhostNet-Investigating-a-Cyber-Espionage-Network; and Information Warfare Monitor and Shadowserver Foundation, Shadows in the Cloud: Investigating Cyber Espionage 2.0, April 5, 2010, http://www.infowar-monitor.net/2010/04/shadows-in-the-cloud-an-investigation-into-cyber-espionage-2-0/. John Markoff reported on our investigations in, "Vast Spy System Loots Computers in 103 Countries," *New York Times*, March 28, 2009, http://www.nytimes.com/2009/03/29/technology/29spy.html?pagewanted=all&_r=0; and together with David Barboza in, "Researchers Trace Data Theft to Intruders in China," *New York Times*, April 5, 2010, http://www.nytimes.com/2010/04/06/science/06cyber.html?pagewanted=all.

23   **a huge compromise of American military and intelligence agencies:** For more on "Titan Rain," see James A. Lewis, *Computer Espionage, Titan Rain and China, Center for Strategic and International Studies*, December 2005, http://csis.org/files/media/csis/pubs/051214_china_titan_rain.pdf.

25   **"Who done it?":** Useful primers on the difficulties of attributing the sources of cyber attacks can be found in David D. Clark and Susan Landau, "Untangling Attribution," and W. Earl Boebert, "A Survey of Challenges in Attribution," both of which can be found in *Proceedings of a Workshop on Deterring Cyberattacks: Informing Strategies and Developing Options for U.S. Policy*, 2010, http://www.nap.edu/catalog/12997.html.

2: FILTERS AND CHOKEPOINTS

30   **What is cyberspace?:** Canadian science fiction author William Gibson is credited with coining the term *cyberspace* in his short story "Burning Chrome" (New York: HarperCollins, 2003), and popularizing it in his novel *Neuromancer* (New York: Ace, 1984). Although *cyberspace* and

Internet are often used interchangeably, they are not the same. The Internet is a global network of computer networks configured to operate according to a common protocol of intercommunications (the TCP/IP protocol). Cyberspace is broader and includes the entire domain of global communications, including (but not limited to) the Internet.

31 **Every device we use to connect to the Internet:** Attempts to control cyberspace often start with interventions in the physical infrastructure, specifically at key chokepoints. This has been documented in Ronald Deibert, John Palfrey, Rafal Rohozinski, and Jonathan Zittrain, eds., *Access Denied: The Practice and Policy of Global Internet Filtering* (Cambridge: MIT Press, 2008); Mark Newman, *Networks: An Introduction* (New York: Oxford University Press, 2010); and David D. Clark, "Control Point Analysis" (Paper presented at the 2012 TPRC, 40th Research Conference on Communication, Information and Internet Policy, Arlington, Virginia, September 21–23, 2012), available at: http://dx.doi.org/10.2139/ssrn.2032124.

32 **Much of the software that operates cyberspace is "closed," or proprietary:** On studies of the security of closed- and open-source operating systems, see Kishen Iyengar, M.K. Raja, and Vishal Sachdev, "A Security Comparison of Open-Source and Closed-Source Operating Systems" (Proceedings of South West Decision Sciences Institute's Thirty-eighth Annual Conference, San Diego, CA, 2007), http://www.swdsi.org/swdsi07/2007_proceedings/papers/236.pdf; and Jim Rapoza, "eWeek Labs: Open Source Quicker at Fixing Flaws," eWeek, September 30, 2012, http://www.eweek.com/c/a/Application-Development/eWeek-Labs-Open-Source-Quicker-at-Fixing-Flaws/. An in-depth analysis of the political economy of open-source software can be found in Steven Weber, *The Success of Open Source* (Boston: Harvard University Press, 2008).

33 **In 2010, while mapping for its popular Street View service:** For more on Google's Street View wifi controversy, see David Kravets, "An Intentional Mistake: The Anatomy of Google's Wi-Fi Sniffing Debacle," *Wired*, May 2, 2012, http://www.wired.com/threatlevel/2012/05/google-wifi-fcc-investigation.

34 **In 2012, Cisco provided updates to its popular Linksys:** Cisco's updates are detailed in Joel Hruska, "Cisco's Cloud Vision: Mandatory, Monetized, and Killed at Their Discretion," *Extreme Tech*, July 2, 2012,

http://www.extremetech.com/computing/132142-ciscos-cloud-vision-mandatory-monetized-and-killed-at-their-discretion.

34   **in 2012, a cyber security researcher named Mark Wuergler:** Mark Wuergler's research on the exposure of MAC addresses in Apple devices has been documented in Dan Goodin, "Loose-Lipped iPhones Top the List of Smartphones Exploited by Hacker," *Ars Technica*, March 16, 2012, http://arstechnica.com/apple/2012/03/loose-lipped-iphones-top-the-list-of-smartphones-exploited-by-hacker/.

39   **In 2012, ONI discovered that users in Oman:** The OpenNet Initiative documented its findings on upstream filtering affecting Omani ISP Omantel in Citizen Lab, "Routing Gone Wild: Documenting Upstream Filtering in Oman via India," 2012, https://citizenlab.org/2012/07/routing-gone-wild.

39   **In 2005, ONI found that when the Canadian ISP Telus blocked:** The OpenNet Initiative documented its findings on collateral filtering by Telus in "Telus Blocks Consumer Access to Labour Union Web Site and Filters an Additional 766 Unrelated Sites," *OpenNet Initiative*, August 2, 2005, http://opennet.net/bulletins/010/.

40   **In 2008, the Pakistan Ministry of Information ordered Pakistan Telecom:** Pakistan's 2008 collateral filtering of YouTube is documented in Martin A. Brown, "Pakistan Hijacks YouTube," *Renesys*, February 24, 2008, http://www.renesys.com/blog/2008/02/pakistan-hijacks-you-tube-1.shtml.

40   **there is a deeper layer of control:** For IXPs, see "Internet Exchange Map," *TeleGeography*, http://www.telegeography.com/telecom-resources/internet-exchange-map/index.html; Brice Augustin, Balachander Krishnamurthy, and Walter Willinger, "IXPs: Mapped?", Internet Measurement Conference, November 2009, http://www-rp.lip6.fr/~augustin/ixp/imc2009.pdf. The University of Toronto's IXmaps is a tool that allows researchers to examine the route(s) that data packets take to travel across North America. The tool can be found at: http://www.ixmaps.ca.

41   **In 2002, Mark Klein, a twenty-year veteran technician with AT&T:** Mark Klein's personal statement about discovering the AT&T eavesdropping

facility was published in "Wiretap Whistle-Blower's Account," *Wired*, April 7, 2006, http://www.wired.com/science/discoveries/news/2006/04/70621. The Electronic Frontier Foundation has documented the case at "Hepting v. AT&T," https://www.eff.org/cases/hepting.

44 **In a May 2012 article:** Sam Biddle outlines the physical elements of the Internet that could be targeted in "How to Destroy the Internet," *Gizmodo*, May 23, 2012, http://ca.gizmodo.com/5912383/how-to-destroy-the-internet.

45 **The cause of the severed cables is unknown:** The 2008 severing of cable systems in the Mediterranean Sea is detailed in Asma Ali Zain, "Cable Damage Hits One Million Internet Users in U.A.E.," *Khaleej Times*, February 4, 2008, http://www.khaleejtimes.com/DisplayArticleNew.asp?section=theuae&xfile=data/theuae/2008/february/theuae_february121.xml. See also Andrew Blum, *Tubes: A Journey to the Center of the Internet* (New York: HarperCollins, 2012).

46 **a defunct and wayward Russian satellite:** The 2009 satellite collision is detailed in "Satellite Collision Leaves Significant Debris Clouds," *Orbital Debris Quarterly News* 13, no.2 (2009). The Kessler Syndrome is discussed in detail in Burton G. Cour-Palais and Donald J. Kessler, "Collision Frequency of Artificial Satellites: The Creation of a Debris Belt," *Journal of Geophysical Research* 83 (1978): 2637–2646. See also Daniel H. Deudney, "High Impacts: Asteroidal Utilization, Collision Avoidance, and the Outer Space Regime," in ed. W. Henry Lambright, *Space Policy in the Twenty-First Century* (Baltimore: Johns Hopkins University Press, 2003).

47 **Space is also an arena within which state intelligence agencies:** The literature on American and Soviet space assets developed during the Cold War is sparse because of secrecy. Some important exceptions are Jeffrey Richelson, *America's Space Sentinels* (Lawrence: University of Kansas Press, 1999); William E. Burrows, *Deep Black: Space Espionage and National Security* (New York: Random House, 1986); James Bamford, *The Shadow Factory: The Ultra-Secret NSA from 9/11 to the Eavesdropping on America* (New York: Doubleday, 2008); and Ronald J. Deibert, "Unfettered Observation: The Politics of Earth Monitoring From Space," in ed. W. Henry Lambright, *Space Policy in the Twenty-First Century*.

50 **Big Data: They Reap What We Sow:** On big data, see danah boyd and Kate Crawford, "Critical Questions for Big Data: Provocations for a Cultural, Technological, and Scholarly Phenomenon," *Information, Communication, & Society* 15, no.5 (2012): 662–679; and David Bollier, *The Promise and Peril of Big Data*, Aspen: The Aspen Institute, 2010, http://www.aspeninstitute.org/sites/default/files/content/docs/pubs/The_Promise_and_Peril_of_Big_Data.pdf.

50 **Malte Spitz had virtually every moment of his life tracked:** The Malte Spitz timeline can be found in Malte Spitz, "Six Months of My Life in 35,000 Records," http://www.malte-spitz.de/blog/4103927.html.

52 **IBM predicts that in 2013, we will be producing five exabytes:** IBM explains big data in "What is Big Data?," IBM, http://www-01.ibm.com/software/data/bigdata/. See also "200 Million Tweets Per Day," Twitter, June 30, 2011, http://blog.twitter.com/2011/06/200-million-tweets-per-day.html. In "Gigatweet/Counter," *GigaTweet*, November 6, 2011, http://gigatweeter.com/counter, the tweet-tracking service determined that as of November 6, 2011, 29,700,500,268 tweets had been created. GigaTweet's counter has since stopped due to "technical changes in the way Twitter generates their tweet IDs." See also Alex Hudson, "The Age of Information Overload," BBC, August 14, 2012, http://news.bbc.co.uk/2/hi/programmes/click_online/9742180.stm.

52 **mobile data traffic more than doubled:** Cisco's latest global mobile data traffic forecast can be found in "Cisco Visual Networking Index: Global Mobile Data Traffic Forecast Update, 2011–2016," *Cisco*, 2012, http://www.cisco.com/en/US/solutions/collateral/ns341/ns525/ns537/ns705/ns827/white_paper_c11-520862.pdf.

54 **Simply by collating the number, location, and frequency of search queries:** More on Google Flu Trends can be found at "Tracking Flu Trends," *Google Official Blog,* November 11, 2008, http://googleblog.blogspot.ca/2008/11/tracking-flu-trends.html.

55 **Researchers at Stanford University are testing an app:** The M-Maji water app has been discussed in Sarina A. Beges, "In Kenyan Slum, Mobile Phones Pinpoint Better Water," Program on Liberation Technology, October 26, 2012, http://liberationtechnology.stanford.edu/

news/in_kenyan_slum_mobile_phones_pinpoint_better_
water_20121026/.

56 **The big-data market stood at just over $5 billion:** David Floyer,
Jeff Kelly, and David Vellante list IBM, Intel, and HP as the current
big-data market leaders (by revenue) in "Big Data Market Size and
Vendor Revenues," Wikibon, May 29, 2012, http://wikibon.org/wiki/v/
Big_Data_Market_Size_and_Vendor_Revenues.

56 **Customized dating and vacation ads:** Claude Castelluccia,
Mohamed-Ali Kaafar, and Minh-Dung Tran examine the privacy con-
cerns generated by the practice of tracking users' online behaviours in
"Betrayed By Your Ads!: Reconstructing User Profiles from Targeted Ads,"
(Paper presented at the 12th Privacy Enhancing Technologies Symposium,
Vigo, Spain, July 10–13, 2012), http://dl.acm.org/citation.cfm?id=2359017.

57 **A tracking-awareness project:** More on the Collusion tool is
provided on Mozilla's website at "Introducing Collusion: Discover Who's
Tracking You Online," http://www.mozilla.org/en-US/collusion/. The
Wall Street Journal has documented the widespread use of tracking
technology and what this type of surveillance means for consumers and
society in its "What They Know" series: http://online.wsj.com/public/
page/what-they-know-2010.html.

58 **The small print included with many applications and/or service
contracts:** Tom Kelly investigates fourteen popular mobile apps and
documents the type of personal information and data the apps can access
in "Free Apps 'Can Spy on Texts and Calls': Smartphone Users Warned
of Privacy Dangers," *Daily Mail*, February 27, 2012, http://www.
dailymail.co.uk/sciencetech/article-2106627/Internet-firms-access-
texts-emails-pictures-spying-smartphone-apps.html.

59 **Facebook's European headquarters is in Dublin, Ireland:** Ireland
has strict privacy laws to which Facebook must adhere, as reported in
Kashmir Hill, "Max Schrems: The Austrian Thorn In Facebook's Side,"
Forbes, February 7, 2012, http://www.forbes.com/sites/kash-
mirhill/2012/02/07/the-austrian-thorn-in-facebooks-side/.

59 **Over the years, Facebook's default privacy settings:** A timeline
documenting changes to Facebook's privacy policy can be found in Kurt

Opsahl, "Facebook's Eroding Privacy Policy: A Timeline," Electronic
Frontier Foundation, April 28, 2010, https://www.eff.org/deep-
links/2010/04/facebook-timeline.

60 **caught uploading members' mobile phone contacts:** In 2012,
many major app companies were sued in a class action lawsuit for selling
mobile apps that uploaded users' address book data without their
knowledge or consent. Companies included Path, Twitter, Apple.
Facebook, Beluga, Yelp, Burbn, Instagram, Foursquare, Gowalla,
Foodspotting, LinkedIn, Electronic Arts, Kik Interactive, and more. See
"Tons of Companies Sued in Class Action Lawsuit over Uploading
Phone Addressbooks," *Tech Dirt*, March 20, 2012, http://www.techdirt.
com/articles/20120316/00561518126/tons-companies-sued-class-action-
lawsuit-over-uploading-phone-addressbooks.shtml. See also Julia Angwin
and Jeremy Singer-Vine, "Selling You on Facebook," *Wall Street Journal*,
April 7, 2012, http://online.wsj.com/article/SB1000142405270230330250
4577327744009046230.html?mod=WSJ_WhatTheyKnowPrivacy_
MIDDLETopMiniLeadStory; Nick Bilton and Nicole Perlroth,
"Mobile Apps Take Data without Permission," *New York Times*,
February 15, 2012, http://bits.blogs.nytimes.com/2012/02/15/google-
and-mobile-apps-take-data-books-without-permission/; "Now Twitter
Admits 'Harvesting' Users' Phone Contacts Without Telling the
Owners as Apple Announces Crackdown," *Daily Mail*, February 16, 2012,
http://www.dailymail.co.uk/sciencetech/article-2101934/Apple-moves-
stop-Facebook-Twitter-accessing-iPhone-users-address-books-
permission.html; "Statement of Justin Brookman," Before the Senate
Judiciary Committee, Subcommittee on Privacy, Technology, and the
Law, Hearing on Protecting Mobile Privacy: Your Smartphones, Tablets,
Cell Phones, and Your Privacy, May 10, 2011, https://www.cdt.org/files/
pdfs/20110510_mobile_privacy.pdf; and Lito Cruz, Andre Olober, and
Kristopher Welsh, "The Danger of Big Data: Social Media as
Computational Social Science," *First Monday* 17, no.7 (2012), http://
www.firstmonday.org/htbin/cgiwrap/bin/ojs/index.php/fm/article/
view/3993/3269.

61 **U.S. Federal Trade Commission found that Facebook had
engaged in:** The FTC accused Facebook of deceiving "consumers by
telling them they could keep their information on Facebook private, and
then repeatedly allowing it to be shared and made public." This was
documented in Dominic Rushe, "Facebook Reaches Deal with FTC

Over 'Unfair and Deceptive' Privacy Claims," *Guardian*, November 29, 2011, http://www.guardian.co.uk/technology/2011/nov/29/facebook-ftc-privacy-settlement.

62   **freedom of information request to find out more about the secret agreement:** Electronic Privacy Information Center's (EPIC) executive director, Marc Rotenberg, believed that the Google–NSA agreement covered more than just the Google breach, and that both Google and the NSA were in talks before Google found out that it had been compromised in a computer attack. See Adam Gabbatt, "Google Teams Up with National Security Agency to Tackle Cyber Attacks," *Guardian*, February 5, 2010, http://www.guardian.co.uk/technology/2010/feb/05/google-national-security-agency-cyber-attack; and Jaikumar Vijayan, "Google Taps NSA to Safeguard Its Data," *Computer World*, February 4, 2010, http://www.pcworld.com/article/188557/google_taps_nsa_to_safeguard_data.html.

62   **Network operators and service providers vary:** See "Retention Periods of Major Cellular Service Providers," United States Department of Justice, http://www.wired.com/images_blogs/threatlevel/2011/09/retentionpolicy.pdf.

63   **Polish NGO Panoptykon found that Polish authorities:** Panoptykon discusses the effects of Poland's data retention policies in "How Many Times Did the State Authorities Reach Out for Our Private Telecommunications Data in 2011? We Publish the Latest Research," *Panoptykon*, March 4, 2012, http://panoptykon.org/wiadom-osc/how-many-times-did-state-authorities-reach-out-our-private-telecommunications-data-2011-we.

64   **find a way to integrate as much data as possible:** The Total Information Awareness system is discussed in Shane Harris, *The Watchers: The Rise of America's Surveillance State* (New York: The Penguin Group, 2010). The *Washington Post* has been documenting the national security buildup in the United States that has occurred since 9/11 in "Top Secret America," blog, http://projects.washingtonpost.com/top-secret-america.

66   **As former CIA director David Petraeus explained:** Petraeus's remarks at the 2012 In-Q-Tel CEO Summit are available at "Remarks by Director David H. Petraeus at In-Q-Tel CEO Summit," Central

Intelligence Agency, March 1, 2012, https://www.cia.gov/news-information/speeches-testimony/2012-speeches-testimony/in-q-tel-summit-remarks.html.

67   **In 2012, the Hamburg Commissioner for Data Protection and Freedom of Information:** German data protection officials accused Facebook of "illegally compiling a vast photo database of users without their consent," and demanded that Facebook destroy all archives of files based on facial recognition technology. See Violet Blue, "Why You Should Be Worried About Facial-Recognition Technology," CNET, August 29, 2012, http://news.cnet.com/8301-1023_3-57502284-93/why-you-should-be-worried-about-facial-recognition-technology/.

4: THE CHINA SYNDROME

72   **"In China, the Internet came with choke points built in.":** The OpenNet Initiative has documented Chinese cyberspace controls in "China," in *Access Contested: Security, Identity, and Resistance in Asian Cyberspace*, eds. Ronald Deibert, John Palfrey, Rafal Rohozinski, and Jonathan Zittrain (Cambridge: MIT Press, 2012), 271–298. See also Milton Mueller, "China and Global Internet Governance: A Tiger by the Tail," in eds. Deibert et al., *Access Contested*, 177–194; and Greg Walton, *China's Golden Shield*, International Centre for Human Rights and Democratic Development, 2001.

73   **Contrary to the principles of network neutrality:** *Network neutrality* is a term coined by Tim Wu. For Wu, "Network neutrality is best defined as a network design principle. The idea is that a maximally useful public information network aspires to treat all content, sites, and platforms equally." See Tim Wu "Network Neutrality FAQ," http://timwu.org/network_neutrality.html and "Network Neutrality, Broadband Discrimination," *Journal of Telecommunications and High Technology Law* 2 (2003). See also Lawrence Lessig and Robert W. McChesney, "No Tolls on the Internet," *Washington Post*, June 8, 2006, http://www.washingtonpost.com/wp-dyn/content/article/2006/06/07/AR2006060702108.html; and Milton Mueller, *Net Neutrality as Global Principle for Internet Governance* (Syracuse: Internet Governance Project, 2007).

74   **The Chinese version of Skype:** The TOM-Skype investigation is documented in Nart Villeneuve, "Breaching Trust: An Analysis of

Surveillance and Security Practices on China's TOM-Skype Platform," Information Warfare Monitor, September 2009, http://www.infowar-monitor.net/2009/09/breaching-trust-an-analysis-of-surveillance-and-security-practices-on-china's-tom-skype-platform/. See also John Markoff, "Surveillance of Skype Messages Found in China," *New York Times*, October 1, 2008, http://www.nytimes.com/2008/10/02/technology/internet/02skype.html?pagewanted=all.

Years after the release of the Citizen Lab's TOM-Skype research, researchers from the University of New Mexico found the exact same content-filtering and interception system on TOM-Skype. Their research is documented in Jedidiah R. Crandall, Jeffrey Knockel, and Jared Saia, "Three Researchers, Five Conjectures: An Empirical Analysis of TOM-Skype Censorship and Surveillance (Paper presented at the USENIX Workshop on Free and Open Communications on the Internet, San Francisco, California, August 2011), available at: http://www.cs.unm.edu/~crandall/foci11knockel.pdf. Citizen Lab and UNM are now working together on a study of several Chinese-marketed chat clients and will publish our results in 2013.

75 **Researchers at Cambridge University, for instance, once demonstrated:** See Richard Clayton, Steven J. Murdoch, Robert N.M. Watson, "Ignoring the Great Firewall of China," *Journal of Law and Policy for the Information Society* 3, no.2 (2007). Psiphon was invented in the Citizen Lab, and released in December 2006 at an event called Protect the Net. The project was funded by the Open Society Institute as part of the Citizen Lab's CiviSec Project. Psiphon was spun out of the University of Toronto as an independent Canadian company. Read more about Psiphon at http://psiphon.ca. Karl Kathuria, in *Casting a Wider Net* (a study undertaken in 2011 while a Citizen Lab/Canada Centre visiting fellow), combined ONI, Psiphon, and BBC media data to develop policies for global broadcasters whose content is filtered in censored jurisdictions: http://munkschool.utoronto.ca/downloads/casting.pdf.

75 **Code words, metaphors, neologisms:** Xiao Qiang, editor of the *China Digital Times*, has been compiling a glossary of terms used by creative Chinese netizens to bypass China's online censors in "Grass-Mud Horse Lexicon," *China Digital Times*, http://chinadigitaltimes.net/space/Introduction_to_the_Grass-Mud_Horse_Lexicon.

76 **Often ignored is the connection between China's domestic controls and the international dimensions of its cyberspace strategy:** Masashi Crete-Nishihata and I examine the international and global mechanisms that facilitate the growth and spread of cyberspace controls in "Global Governance and the Spread of Cyberspace Controls," *Global Governance: A Review of Multilateralism and International Organizations* 18, no. 3 (2012): 339–361.

76 **Evidence of GhostNet-like compromises now surface almost weekly:** Jameson Berkow reported on the Nortel breach in "Nortel Hacked to Pieces," *Financial Post*, February 25, 2012, http://business. financialpost.com/2012/02/25/nortel-hacked-to-pieces.

78 **It's unlikely that China would benefit in an armed conflict:** On China's military strategy, see Timothy L. Thomas, *Dragon Bytes: Chinese Information-War Theory and Practice* (Fort Leavenworth: Foreign Military Studies Office, 2004). See also U.S.-China Economic and Security Review Commission, 2012 Report to Congress, 147–169; and Desmond Ball, "China's Cyber Warfare Capabilities," *Security Challenges* 7, iss. 2 (2011): 83–103.

79 **Part of China's international strategy revolves around setting:** On China's technology industry development, see David Chen, Stephen Schlaikjer, and Micah Springut, "China's Program for Science and Technology Modernization: Implications for American Competitiveness," January 2011, http://www.uscc.gov/researchpapers/2011/USCC_ REPORT_China's_Program_forScience_and_Technology_Modernization. pdf; and Steven P. Bucci and Derek Scissors, "China Cyber Threat: Huawei and American Policy Toward Chinese Companies," *Heritage*, October 23, 2012, http://www.heritage.org/research/reports/2012/10/china-cyber-threat-huawei-and-american-policy-toward-chinese-companies. By the end of 2012, a fifth of all computers in the world will be manufactured in Chengdu. See Ambrose Evans-Pritchard, "Hi-tech Expansion Drives China's Second Boom in the Hinterland," *Telegraph*, November 25, 2012, http://www.telegraph.co.uk/finance/comment/9701910/Hi-tech-expansion-drives-Chinas-second-boom-in-the-hinterland.html.

80 **a regional ... security alliance called the Shanghai Cooperation Organization:** An overview of the Shanghai Cooperation Organization can be found in Andrew Scheineson, "The Shanghai Cooperation

Organization," *Council on Foreign Relations*, March 24, 2009, http://www.
cfr.org/publication/10883/shanghai_cooperation_organization.html;
Thomas Ambrosio, "Catching the 'Shanghai Spirit': How the Shanghai
Cooperation Organization Promotes Authoritarian Norms in Central
Asia," *Europe-Asia Studies* 60, no.8 (2008): 1321–1344; and *Human Rights
in China, Counter-Terrorism and Human Rights: The Impact of the Shanghai
Cooperation Organization* (New York: Human Rights in China, 2011),
http://www.hrichina.org/research-and-publications/reports/sco.

5: THE NEXT BILLION DIGITAL NATIVES

82 **Somalia has not had a properly functioning government since
1991:** An extensive look at Somalia's thriving telecommunications sector
is available in Bob Feldman, "Somalia: Amidst the Rubble, a Vibrant
Telecommunications Infrastructure," *Review of African Political Economy*
34, no.113 (2007): 565–572; Sarah Childress and Abdinasir Mohamed,
"Telecom Firms Thrive in Somalia Despite War, Shattered Economy,"
Wall Street Journal, May 11, 2010, http://online.wsj.com/article/SB10001
424052748704608104575220570113266984.html; Joseph Winter,
"Telecoms Thriving in Lawless Somalia," BBC, November 19, 2004,
http://news.bbc.co.uk/2/hi/africa/4020259.stm; and Abdinasir
Mohamed and Sarah Childress, "Telecom Firms Thrive in Somalia
Despite War, Shattered Economy," *Wall Street Journal*, May 11, 2010,
http://online.wsj.com/article/SB10001424052748704608104575220570011
3266984.html.

85 **Somali cab drivers, nurses, teachers, engineers:** Somalia is the
fourth most remittance-dependent country in the world. See Mohamed
Aden Hassan and Caitlin Chalmers, "UK Somali Remittances Survey,"
Department for International Development, May 2008, http://www.
diaspora-centre.org/DOCS/UK_Somali_Remittan.pdf.

88 **According to the *Arab Social Media Report* series:** This series can be
found at Dubai School of Government, "Social Media in the Arab World:
Influencing Societal and Cultural Change?," *Arab Social Media Report* 2, no.1
(2012), http://www.arabsocialmediareport.com/UserManagement/PDF/
ASMR%204%20updated%2029%2008%2012.pdf.

89 **The combination of youth, unemployment, and radicalism:** On
youth unemployment in the Arab Spring, see Gwyn Morgan, "Youth

Unemployment the Kindling that Fuels Unrest," *Globe and Mail*, September 10, 2012, http://m.theglobeandmail.com/report-on-business/rob-commentary/youth-unemployment-the-kindling-that-fuels-unrest/article4199652/?service=mobile.

89 **The fastest growth rates are occurring among the world's failed and most fragile states:** In the ITU's 2009 Information Society Statistical Profiles, the ten countries that saw the fastest Internet user growth rates (calculated in terms of compounded annual growth rates) over five years were Afghanistan, Myanmar, Vietnam, Albania, Uganda, Nigeria, Liberia, Sudan, Morocco, and D.R. Congo. Uganda, Nigeria, Liberia, and D.R. Congo were ranked as having low human development on the UN's 2008 Human Development Index, with no available data for Afghanistan, which at the time was ranked seventh on the Fund for Peace's Failed States Index. The growth rates for Afghanistan, Myanmar, and Vietnam were derived from 2002 to 2007 ITU figures, while 2003 to 2008 figures were used for the rest. The International Telecommunications Union's Information Society Statistical Profiles are available at: "Information Society Statistical Profiles 2009: Africa," 2010, http://www.itu.int/ITU-D/ict/material/ISSP09-AFR_final-en.pdf; "Information Society Statistical Profiles 2009: Europe v.1.01," 2010, http://www.itu.int/dms_pub/itu-d/opb/ind/D-IND-RPM.EUR-2009-R1-PDF-E.pdf; "Information Society Statistical Profiles 2009: Europe," 2010, http://www.itu.int/dms_pub/itu-d/opb/ind/D-IND-RPM.EUR-2009-R1-PDF-E.pdf; "Information Society Statistical Profiles 2009: Americas," 2010, http://www.itu.int/dms_pub/itu-d/opb/ind/D-IND-RPM.AM-2009-E09-PDF-E.pdf; and "Information Society Statistical Profiles 2009: Asia and the Pacific," 2010, http://www.itu.int/dms_pub/itu-d/opb/ind/D-IND-RPM.AP-2009-R1-PDF-E.pdf. See also "Failed States Index 2008," Fund for Peace, http://www.fundforpeace.org/global/?q=fsi-grid2008.

91 **Whereas in other parts of the world cyberspace controls:** Rafal Rohozinski and I discuss cyberspace controls in Russia and the Commonwealth of Independent States in "Control and Subversion in Russian Cyberspace," in *Access Controlled: The Shaping of Power, Rights, and Rule in Cyberspace*, eds. Ronald Deibert, John Palfrey, Rafal Rohozinski, and Jonathan Zittrain (Cambridge: MIT Press, 2010): 15-34.

92 **Each country in the global South and East:** India's "Information Technology (Intermediaries Guidelines) Rules of 2011," available at http://

www.mit.gov.in/sites/upload_files/dit/files/GSR314E_10511%281%29.pdf, place extraordinary policing responsibilities on ISPs and other services that operate in cyberspace. India's 2008 Information Technology Act, available at: http://www.mit.gov.in/sites/upload_files/dit/files/downloads/itact2000/it_amendment_act2008.pdf, gives the government the power to block, intercept, monitor, or decrypt any information in the interest of sovereignty, integrity, defence, or security of India. See Amol Sharma, "India Court Adjourns Google-Facebook Case Until August," *Wall Street Journal*, May 3, 2012, http://online.wsj.com/article/SB100014240527023047466045773817 91461076660.html; and Jonah Force Hill, "India: The New Front Line in the Global Struggle for Internet Freedom," *The Atlantic*, June 7, 2012, http://www.theatlantic.com/international/archive/2012/06/india-the-new-front-line-in-the-global-struggle-for-internet-freedom/258237.

94 **India has also waged a persistent campaign:** In October 2011, it was reported that RIM had set up a facility in Mumbai to help the Indian government carry out lawful surveillance. See "RIM Sets Up Facility to Help Indian Government with Lawful Surveillance," *Toronto Star*, October 28, 2011, http://www.thestar.com/business/article/1077575--rim-sets-up-facility-to-help-indian-government-with-lawful-surveillance?bn=1. I wrote about the travails of RIM in "Cyberspace Confidential," *Globe and Mail*, August 6, 2010, http://www.theglobeandmail.com/commentary/cyberspace-confidential/article1241035/?page=all.

95 **Meanwhile, the Indian government banned all mass text messaging** Regarding the Indian government's banning mass texting for two weeks, see Dean Nelson, "India Bans Mass Text Messages to Stem Panic Among Minorities," *Telegraph*, August 17, 2012, http://www.telegraph.co.uk/news/worldnews/asia/india/9482890/India-bans-mass-text-messages-to-stem-panic-among-minorities.html.

95 **in an attempt to prevent the sale and distribution of cloned and pirated mobile phones:** On the Communications Commission of Kenya cutting off 1.89 million mobile phones, see Winfred Kagwe, "Kenya: 1.9 Million 'Fake' Phones Shut," *All Africa*, October 2, 2012, http://allafrica.com/stories/201210020512.html.

95 **In 2010, Turkey ordered ISPs to block access to YouTube:** YouTube was banned by a 2007 court decision after videos accusing Mustafa Kemal Ataturk, the first president of Turkey, of being

homosexual. In an attempt to enforce this ban, Turkey also blocked more than thirty Google services. See Justin Vela, "Turkey Blocks Google Sites – Accidentally?" *AOL News*, June 9, 2010, http://www.aolnews. com/2010/06/09/turkey-blocks-google-sites-accidently/.

95 **South Korea put in place regulations to block several dozen North Korean websites:** The OpenNet Initiative reported on South Korean collateral filtering in "Collateral Blocking: Filtering by South Korean Government of Pro-North Korean Websites," January 31, 2005, http://opennet.net/bulletins/009/.

95 **In his 2010 report:** Helmi Noman discusses faith-based Internet censorship in majority Muslim countries in "In the Name of God: Faith-based Internet Censorship in Majority Muslim Countries," OpenNet Initiative, August 1, 2011, http://opennet.net/sites/opennet. net/files/ONI_NameofGod_1_08_2011.pdf.

97 **In the Far East, the same pattern is emerging:** The OpenNet Initiative has documented Thailand's cyberspace controls in "Thailand," in *Access Contested*, eds. Ronald Deibert et al., 271–298. The case of Chiranuch Premchaiporn is detailed in James Hookway, "Conviction in Thailand Worries Web Users," *Wall Street Journal*, May 30, 2012, http://online.wsj. com/article/SB10001424052702303674004577435373324265632.html.

99 **The cartels have also shown a ruthless ability:** The murders of bloggers by drug cartels in Mexico has been profiled in "Mexican Drug Gang Beheads ANOTHER Blogger and Dumps Body and Severed Head in Street with Bloody Warning Note," *Daily Mail Online*, November 10, 2011, http://www.dailymail.co.uk/news/ article-2060057/Blogger-beheaded-Mexican-gang-left-note-warning-snitch.html; and "Killings Grow More Gruesome as Mexican Drug Cartels Try to Out-shock," *The National*, October 10, 2011, http://www. thenational.ae/news/world/americas/killings-grow-more-gruesome-as-mexican-drug-cartels-try-to-out-shock. The Citizen Lab's Luis Horacio Najera is an exiled journalist from Mexico and winner of the 2010 International Press Freedom Award. He has been undertaking extensive research on the use of information and communication technologies by Latin American drug cartels and will be publishing his findings in 2013 as a Citizen Lab report.

103 **Google's ongoing acrimonious relationship with China:** Google's announcement of the two policies is available on its blog at "Security Warnings for Suspected State-Sponsored Attacks," June 5, 2012, http://googleonlinesecurity.blogspot.ca/2012/06/security-warnings-for-suspected-state.html; and "Better Search in Mainland China," *Inside Search: The Official Google Search Blog*, May 31, 2012, http://insidesearch.blogspot.sg/2012/05/better-search-in-mainland-china.html.

104 **"corporate sovereignty":** Rebecca MacKinnon, *Consent of the Networked: The Worldwide Struggle for Internet Freedom* (New York: Basics Books, 2012). In *The Master Switch: The Rise and Fall of Information Empires* (New York: Random House, 2010), Tim Wu shows how all previous innovations of the information industry have followed a single path from being open and widely accessible to being dominated by a single corporation or cartel, and warns that the Internet may one day also follow this path of development.

106 **Its vigorous opposition to the SOPA and PIPA bills:** The Stop Online Piracy Act (SOPA) and the Protect IP Act (PIPA) aim to curtail online copyright violations by granting the U.S. government new tools and powers to block users' access to websites that sell copyright-infringing or counterfeit goods.

107 **"unprecedented synthesis of corporate and public spaces":** Steve Coll's *New Yorker* essay "Leaving Facebookistan" (May 24, 2012) is available at, http://www.newyorker.com/online/blogs/comment/2012/05/leaving-facebookistan.html.

107 **social media are less like town squares:** On private policing of online content, see Jillian C. York, "Policing Content in the Quasi-Public Sphere," OpenNet Initiative, September 2010, http://opennet.net/policing-content-quasi-public-sphere.

109 **they have had to balance the desire to penetrate markets:** On corporate social responsibility, see John Palfrey and Jonathan Zittrain, "Reluctant Gatekeepers: Corporate Ethics on a Filtered Internet," in Deibert et al., eds., *Access Denied*, 103–122. See also Colin M. Maclay, "Protecting Privacy and Expression Online: Can the Global Network Initiative Embrace the Character of the Net?," in *Access Controlled*,

87–108; Ethan Zuckerman, "Intermediary Censorship," in *Access Controlled*, 71–86; and Jonathan Zittrain, "Be Careful What You Ask For: Reconciling a Global Internet and Local Law," in *Who Rules the Net*, eds. Adam Thierer and Clyde Wayne Crews (Washington: Cato Institute, 2003).

110   **The same downloading of responsibilities can be seen in:** A larger discussion of the concerns associated with the Anti-Counterfeiting Trade Agreement is available in Michael Geist, "The Trouble with the Anti-Counterfeiting Trade Agreement (ACTA)," *SAIS Review* 30.2 (2010).

110   **stating that it archives content removal requests:** The Chilling Effects Clearinghouse is a joint project of the Electronic Frontier Foundation and Harvard University, Stanford University, University of California, Berkeley, University of San Francisco, University of Maine, George Washington University Law School, and Santa Clara University School of Law clinics. More on Chilling Effects can be found in, http://www.chillingeffects.org/faq.cgi.

7: POLICING CYBERSPACE: IS THERE AN "OTHER REQUEST" ON THE LINE?

112   **In November 2012, Google released an update:** The Google Transparency Report can be found at http://www.google.com/transparencyreport.

116   **Twitter's report came out immediately:** The Twitter/Malcolm Harris case was profiled in Joseph Ax, "Occupy Wall Street Protester Whose Tweets Were Subpoenaed to Plead Guilty," *Reuters*, December 5, 2012, http://www.reuters.com/article/2012/12/06/us-twitter-occupy-idUSBRE8B504120121206.

117   **The EFF has investigated and ranked eighteen U.S. email, ISP, and cloud storage companies:** In "When the Government Comes Knocking, Who Has Your Back?," https://www.eff.org/pages/who-has-your-back/, the Electronic Frontier Foundation "examined the policies of 18 major Internet companies – including email providers, ISPs, cloud storage providers, and social networking sites – to assess whether they publicly commit to standing with users when the government seeks access to user data." See also Christopher Soghoian, "An End to Privacy Theater: Exposing and Discouraging Corporate Disclosure of User Data

to the Government," *Minnesota Journal of Law, Science & Technology*, 12, no.1 (2011): 191–237.

118 **Thai–American citizen detained:** In August 2011, Anthony Chai, with the support of the World Organization for Human Rights, filed a lawsuit with a central California district court, charging Netfirm of violating (1) Canada's Personal Information Protection and Electronic Documents Act (PIPEDA); (2) the privacy provisions in California's Business and Professions Code; and (3) the Declaration of Rights contained in California's constitution. Chai is suing the company for US$75,000 in restitution and punitive damages. See Matthew Lasar, "Thai Censorship Critic Strikes Back at Snitch Web Host," *Ars Technica*, August 29, 2011, http://arstechnica.com/business/2011/08/thai-dissi-dent-strikes-back-at-snitch-web-host/. The OpenNet Initiative has documented Thailand's cyberspace controls in "Thailand," in *Access Contested*, eds. Ronald Deibert et al., 271–298.

119 **The Chinese government requested information:** Rebecca MacKinnon discusses corporate social responsibility in the case of Yahoo! and Shi Tao in Shi Tao, Yahoo!, and the Lessons for Corporate Social Responsibility, Version 1.0, December 20, 2007, http://rconversation. blogs.com/YahooShiTaoLessons.pdf.

122 **numerous demands by governments to eavesdrop on users:** Christopher Parsons investigates BlackBerry security, and government requests for its decryption keys, in "Decrypting Blackberry Security, Decentralizing the Future," *Technology, Thoughts, and Trinkets*, November 29, 2010, http://www.christopher-parsons.com/blog/technology/ decrypting-blackberry-security-decentralizing-the-future. The Citizen Lab's announcement of the RIM Check project is at "Information Warfare Monitor (Citizen Lab and SecDev Group) Announces RIM Monitoring Project," Information Warfare Monitor, October 21, 2010, http://www. infowar-monitor.net/2010/10/information-warfare-monitor-citizen-lab-and-secdev-group-announces-rim-monitoring-project/.

125 **A June 2012 Human Rights Watch (HRW) report:** See in "In the Name of Security, Counterterrorism Laws Worldwide Since September 11," *Human Rights Watch*, 2012, http://www.hrw.org/sites/default/files/ reports/global0612ForUpload_1.pdf.

126　**ATIS hosts a number of committees and subcomittees:** Ryan Gallagher discusses how networks of telecom companies and international government agencies, such as the Alliance for Telecommunications Industry Solutions (ATIS), are responsible for the harmonization of surveillance laws in "How Governments and Telecom Companies Work Together on Surveillance Laws," *Slate*, August 14, 2012, http://www.slate.com/articles/technology/future_tense/2012/08/how_governments_and_telecom_companies_work_together_on_surveillance_laws_.html.

127　**dozens of governments party to this agreement:** More information on the Council of Europe's Convention on Cybercrime is available in Amalie M. Weber, "The Council of Europe's Convention on Cybercrime," *Berkeley Technology Law Journal* 18, no.1 (2003).

127　**would require ISPs and other telecommunication companies to store:** The proposed Communications Data Bill has been profiled in "UK's Data Communication Bill Faces Tough Criticism," BBC, June 14, 2012, http://www.bbc.com/news/technology-18439226; "Jimmy Wales, Tim Berners-Lee Slam UK's Internet Snooping Plans," *ZDNet*, September 6, 2012, http://www.zdnet.com/uk/jimmy-wales-tim-berners-lee-slam-uks-internet-snooping-plans-7000003829; "UK's Web Monitoring Draft Bill Revealed: What You Need to Know," *ZDNet*, June 14, 2012, http://www.zdnet.com/blog/london/uks-web-monitoring-draft-bill-revealed-what-you-need-to-know/5183; and Mark Townsend, "Security Services to Get More Access to Monitor Emails and Social Media," *Guardian*, July 28, 2012, http://www.guardian.co.uk/technology/2012/jul/28/isecurity-services-emails-social-media.

128　**From documents released under federal access to information laws:** See Christopher Parsons, "Canadian Social Media Surveillance: Today and Tomorrow," *Technology, Thoughts, and Trinkets*, May 28, 2012, http://www.christopher-parsons.com/blog/technology/canadian-social-media-surveillance-today-and-tomorrow/.

8: MEET KOOBFACE: A CYBER CRIME SNAPSHOT

133　**Meet Koobface: A Cyber Crime Snapshot:** Between April and November 2010, the Information Warfare Monitor, led by Nart Villeneuve, conducted an investigation into the operations and monetization strategies of the Koobface botnet. See Nart Villeneuve, "Koobface: Inside a

Crimeware Network," *Information Warfare Monitor*, 2010, http://www.
infowar-monitor.net/reports/iwm-koobface.pdf. Other important studies on Koobface include Jonell Baltazar, Joey Costoya, and Ryan Flores,
"The Real Face of KOOBFACE: The Largest Web 2.0 Botnet Explained,"
TrendWatch, July 2009, http://us.trendmicro.com/imperia/md/content/
us/trendwatch/researchandanalysis/the_real_face_of_koobface_jul2009.
pdf; Jonell Baltazar, Joey Costoya, and Ryan Flores, "The Heart of
KOOBFACE: C&C and Social Network Propagation," *TrendWatch*, October
2009, http://us.trendmicro.com/imperia/md/content/us/trendwatch/
researchandanalysis/the_20heart_20of_20koobface_final_1_.pdf;
Jonell Baltazar, Joey Costoya, and Ryan Flores, "Show Me the
Money! The Monetization of KOOBFACE," *Trend Watch*, November
2009, http://us.trendmicro.com/imperia/md/content/us/trendwatch/
researchandanalysis/koobface_part3_showmethemoney.pdf; and
Jonell Baltazar, "Web 2.0 Botnet Evolution: KOOBFACE Revisited,"
TrendWatch, May 2010, http://us.trendmicro.com/imperia/md/content/
us/trendwatch/researchandanalysis/web_2_0_botnet_evolution_-
_koobface_revisited_may_2010_.pdf. In January 2012, Jan Droemer
and Dirk Kollberg reported on their own detailed investigation of the
Koobface perpetrators in "The Koobface Malware Gang Exposed,"
Sophos Lab, January 2012, http://www.sophos.com/medialibrary/
PDFs/other/sophoskoobfacearticle_rev_na.pdf?dl=true.

139 **Electrons may move at the speed of light, but legal systems
crawl at the speed of bureaucratic institutions:** The lack of
international co-operation around cyber security is discussed in Brian
Krebs, "From (& To) Russia, With Love," *Washington Post*, March 3, 2009,
http://voices.washingtonpost.com/securityfix/2009/03/from_to_russia_
with_love.html. See also Jeremy Kirk, "UK Police Reveal Arrests Over
Zeus Banking Malware," *Computer World*, November 18, 2009, http://
www.computerworld.com/s/article/9141092/UK_police_reveal_arrests_
over_Zeus_banking_malware; and Omar El-Akkad, "Canadian Firm
Helps Disable Massive Botnet," *Globe and Mail*, March 3, 2010, http://
www.globeandmail.com/news/technology/canadian-firm-helps-disable-
massive-botnet/article1488838.

140 **Specialists working for Facebook, Jan Droemer, and other
security researchers:** In January 2012, Facebook outed the identity of
the Koobface perpetrators in "Facebook's Continued Fight Against
Koobface," January 17, 2012, https://www.facebook.com/note.

php?note_id=10150474399670766. See Riva Richmond, "Web Gang Operating in the Open," *New York Times*, January 16, 2012, http://www. nytimes.com/2012/01/17/technology/koobface-gang-that-used-face-book-to-spread-worm-operates-in-the-open.html?pagewanted=1&_r=2&mid=57&ref=technology. Joe Sullivan, Facebook's chief of security, stated: "People who engage in this type of stuff need to know that their name and real identity are going to come out eventually and they're going to get arrested and they're going to be targeted." A week before Facebook released the identities of the Koobface perpetrators, Dancho Danchev independently released the identity of the leader of Koobface, Anton Nikolaevich Korotchenko of St. Petersburg, in "Who's Behind the Koobface Botnet? – An OSINT Analysis," *Dancho Danchev's Blog – Mind Streams of Information Security Knowledge*, January 9, 2012, http://ddanchev.blogspot.ca/2012/01/whos-behind-koobface-botnet-osint.html. The public exposure and the release of the Sophos report led to immediate action by Koobface: its command-and-control servers stopped responding, and the gang started removing traces of themselves off the Net. Facebook's "name-and-shame approach" was criticized by some in the security community for hampering an ongoing criminal investigation and jeopardizing the evidence. See Stefan Tanase, "Was the Koobface Expose the Right Move?," *Threat Post*, January 19, 2012, http://threatpost.com/en_us/blogs/was-koob-face-expose-right-move-011912.

141 **Ever since the Internet emerged from the world of academia:** A detailed look at modern cyber crime can be found in Misha Glenny, *DarkMarket: How Hackers Became the New Mafia* (Toronto: House of Anansi Press Inc, 2011); Misha Glenny, "Dark Market: Cybercrime, Cybercops and You," *Independent*, September 30, 2011, http://www. independent.co.uk/arts-entertainment/books/reviews/dark-market-cybercrime-cybercops-and-you-by-misha-glenny-2362945.html; and Joseph Menn, *Fatal System Error: The Hunt for the New Crime Lords Who Are Bringing Down the Internet* (New York: Public Affairs, 2010).

142 **In Brazil, there is an academy:** Kaspersky Lab's Fabio Assolini writes about a Brazilian cyber-crime school in "A School for Cybercrime: How to Become a Black Hat," *Secure List*, January 17, 2012, http://www. securelist.com/en/blog/208193337/A_School_for_Cybercrime_How_to_Become_a_Black_Hat.

144 **Cyber crime has become one of the world's largest growth businesses:** General Keith Alexander, NSA director and head of U.S. Cyber Command, recently said that cyber crime and cyber espionage accounted for the greatest transfer of wealth in history. See "America's Top Cyberwarrior Says Cyberattacks Cost $250 Billion a Year," *International Business Times*, July 13, 2012, http://www.ibtimes.com/americas-top-cyberwarrior-says-cyberattacks-cost-250-billion-year-722559.

146 **First, a December 27, 2011, breach:** The breaches that occurred in the last week of December 2011 are documented in "Tianya Hacked, 4 Million Passwords Published," *Tech in Asia,* December 26, 2011, http://www.techinasia.com/tianya-hacked-4-million-passwords-published/; and Ken Dilanian, "Hackers Reveal Personal Data of 860,000 Stratfor Subscribers," *Los Angeles Times*, January 4, 2012, http://articles.latimes.com/2012/jan/04/nation/la-na-cyber-theft-20120104.

146 **a particularly malignant backdoor trojan horse:** Poison Ivy is a common backdoor trojan that gives attackers access to and control of an affected machine. Through the use of the Poison Ivy trojan in the Nitro campaign, attackers were able to steal intellectual property from nearly fifty companies, most of them belonging to the chemical industry. See Eric Chien and Gavin O'Gorman, "The Nitro Attacks: Stealing Secrets from the Chemical Industry," Symantec Security Response, http://www.symantec.com/content/en/us/enterprise/media/security_response/whitepapers/the_nitro_attacks.pdf; and "Nitro Attackers Have Some Gall," *Symantec*, December 12, 2011, http://www.symantec.com/connect/blogs/nitro-attackers-have-some-gall.

147 **in 2009, Koobface left a Christmas greeting for security researchers:** The greeting can be found at Dancho Danchev, "The Koobface Gang Wishes the Industry 'Happy Holidays," Dancho Danchev's Blog – Mind Streams of Information Security Knowledge, December 26, 2009, http://ddanchev.blogspot.ca/2009/12/koobface-gang-wishes-industry-happy.html.

9: DIGITALLY ARMED AND DANGEROUS

153 **the SEA boasted about it on their Arabic Facebook page:** The Syrian Electronic Army (SEA) is an open and organized pro-government computer attack group that is actively targeting political opposition and

Western websites. The Citizen Lab does not have concrete evidence linking the SEA to the Assad regime; however, the regime has expressed tacit support for its activities, and has allowed the group to operate with impunity. See Helmi Noman, "The Emergence of Open and Organized Pro-Government Cyber Attacks in the Middle East: The Case of the Syrian Electronic Army," *Information Warfare Monitor*, May 30, 2011, http://www.infowar-monitor.net/2011/05/7349; "Syrian Electronic Army: Disruptive Attacks and Hyped Targets," *Information Warfare Monitor*, June 25, 2011, http://www.infowar-monitor.net/2011/06/syrian-elec-tronic-army-disruptive-attacks-and-hyped-targets/; and "Syrian Electronic Army Defaces 41 Web sites, One UK Government Web site," *Information Warfare Monitor*, June 29, 2011, http://www.infowar-monitor.net/2011/06/syrian-electronic-army-defaces-41-web-sites-one-uk-gov-ernment-web-site.

155 **In February 2012, Anonymous broke into the email server of the Syrian Ministry:** See Barak Ravid, "Bashar Assad Emails Leaked, Tips for ABC Interview Revealed," Haaretz, February 7, 2012, http://www.haaretz.com/print-edition/news/bashar-assad-emails-leaked-tips-for-abc-interview-revealed-1.411445. The role of Telecomix in distributing circumvention tools to Syrian citizens has been profiled in "#OpSyria: When the Internet does not let citizens down," *Reflets*, September 11, 2011, http://reflets.info/opsyria-when-the-internet-does-not-let-citizens-down/.

156 **routers belonging to Blue Coat:** The Citizen Lab reported on the use of Blue Coat in Syria and Burma in "Behind Blue Coat: Investigations of Commercial Filtering in Syria and Burma," November 9, 2011, https://citizenlab.org/2011/11/behind-blue-coat/; and "Behind Blue Coat: An Update from Burma," November 29, 2011, https://citizenlab.org/2011/11/behind-blue-coat-an-update-from-burma/. On October 5, 2011, Telecomix released censorship log files taken from Syrian Blue Coat devices, showing that the Assad regime was using Blue Coat devices to filter and monitor HTTP connections in Syria. See Sari Horwitz, "Syria Using American Software to Censor Internet, Expert Says," *Washington Post*, October 22, 2011, http://www.washingtonpost.com/world/national-security/syria-using-american-software-to-censor-internet-experts-say/2011/10/22/gIQA5mPr7L_story.html. See also Citizen Lab, "Planet Blue Coat: Mapping Censorship and Surveillance Tools," January 15, 2013, https://citizenlab.org/planetbluecoat.

157 **the website of Al-Manar:** Citizen Lab documented the hosting of
Hezbullah and Syrian government websites on servers based in Canada
in "The Canadian Connection: An Investigation of Syrian Government
and Hezbullah Web Hosting in Canada," November 17, 2011, http://
citizenlab.org/wp-content/uploads/2011/11/canadian_connection.pdf;
and "The Canadian Connection: One Year Later," November 14, 2012,
https://citizenlab.org/2012/11/the-canadian-connection-one-year-later/.

159 **reports from inside Syria of phishing attacks:** On phishing attacks
around the Syrian conflict, see Eva Galperin and Morgan Marquis-
Boire, "Syrian Activists Targeted with Facebook Phishing Attack,"
Electronic Frontier Foundation, March 29, 2012, https://www.eff.org/
deeplinks/2012/03/pro-syrian-government-hackers-target-syrian-
activists-facebook-phishing-attack; and Eva Galperin and Morgan
Marquis-Boire, "New Wave of Facebook Phishing Attacks Targets Syrian
Activists," Electronic Frontier Foundation, April 24, 2012, https://www.
eff.org/deeplinks/2012/04/new-wave-facebook-phishing-attacks-
targets-syrian-activists. See also Peter Eckersley, "A Syrian Man-In-The-
Middle Attack Against Facebook," Electronic Frontier Foundation, May
5, 2011, https://www.eff.org/deeplinks/2011/05/syrian-man-middle-
against-facebook; and Jennifer Preston, "Seeking to Disrupt Protesters,
Syria Cracks Down on Social Media," *New York Times*, March 23, 2011,
http://www.nytimes.com/2011/05/23/world/middleeast/23facebook.
html?_r=4. Since March 2012, the Electronic Frontier Foundation has
been collecting and analyzing malware that pro-Syrian-regime hackers
have used to target the Syrian opposition. See "State Sponsored
Malware," Electronic Frontier Foundation, https://www.eff.org/issues/
state-sponsored-malware. The Citizen Lab reported on the targeted
attacks on Syrian dissidents in "Syrian Activists Targeted with BlackShades
Spy Software," June 19, 2012, https://citizenlab.org/2012/06/syrian-
activists-targeted-with-blackshades-spy-software/.

    The Citizen Lab and EFF are developing a joint report on informa-
tion operations in the Syrian conflict, to be published in spring 2013. See
also Nart Villeneuve, "Fake Skype Encryption Software Cloaks
DarkComet Trojan," *Trend Micro Blog*, April 20, 2012, http://blog.
trendmicro.com/fake-skype-encryption-software-cloaks-darkcomet-trojan/.

161 **a new model of "active defence":** The phenomenon of autocratic
regimes successfully wielding information technologies for their own
advantage is discussed in Ronald Deibert and Rafal Rohozinski,

"Liberation vs. Control: The Future of Cyberspace," *Journal of Democracy* 24, no.1 (2010): 43–57. See also Larry Diamond, "Liberation Technology," *Journal of Democracy* 21, no.3 (2010): 69–83; and Evgeny Morozov, *The Net Delusion* (New York: PublicAffairs, 2011).

163 **during parliamentary elections in Kyrgyzstan:** The OpenNet Initiative documented the failure and hacking of Kyrgyz websites during the 2005 parliamentary elections in Kyrgyzstan in "Special Report: Kyrgyzstan Election Monitoring in Kyrgyzstan," *OpenNet Initiative*, February 2005, http://opennet.net/special/kg/.

163 **2006 Belarus presidential elections:** The OpenNet Initiative documented the attacks on opposition websites and Internet failure during the 2006 presidential elections in Belarus in "The Internet and Elections: The 2006 Presidential Elections in Belarus (and Its Implications)," *OpenNet Initiative*, April 2006, http://opennet.net/sites/opennet.net/files/ONI_Belarus_Country_Study.pdf.

164 **As Russian tanks stormed the territory:** The use of information controls during the 2008 Russia–Georgia war is discussed in Masashi Crete-Nishihata, Ronald J. Deibert, and Rafal Rohozinski, "Cyclones in Cyberspace: Information Shaping and Denial in the 2008 Russia–Georgia War," *Security Dialogue* 43.1 (February 2012), 3–24.

165 **downloaded instructions for one of the DDoS tools:** Evgeny Morozov wrote about his experience as a participant in the online Georgia-Russia war in "An Army of Ones and Zeroes: How I Became a Soldier in the Georgia-Russia Cyberwar," *Slate*, August 14, 2008, http://www.slate.com/articles/technology/technology/2008/08/an_army_of_ones_and_zeroes.html.

166 **vexing the Burmese opposition and independent media outlets:** The Citizen Lab's research on DDoS and defacement attacks on Burmese opposition and independent media outlets was documented in Masashi Crete-Nishihata and Nart Villeneuve, "Control and Resistance: Attacks on Burmese Opposition Media," in *Access Contested: Security, Identity, and Resistance in Asian Cyberspace*, eds. Ronald Deibert, John Palfrey, Rafal Rohozinski, and Jonathan Zittrain. (Cambridge: MIT Press, 2012): 154–176.

167 **When the Iranian Cyber Army launched:** In September 2011, it came to light that the DigiNotar Certificate Authority was compromised by a lone Iranian hacker. See Peter Eckersley, Eva Galperin, and Seth Schoen, "A Post Mortem on the Iranian DigiNotar Attack," Electronic Frontier Foundation, September 13, 2011, https://www.eff.org/deeplinks/2011/09/post-mortem-iranian-diginotar-attack.

10 : FANNING THE FLAMES OF CYBER WARFARE

171 **Kaspersky is concerned about anonymity online:** When asked "What's wrong with the design of the Internet?" in a 2009 interview with *ZDNet*, Kaspersky responded: "There's anonymity. Everyone should and must have an identification, or Internet passport. The Internet was designed not for public use, but for American scientists and the U.S. military. That was just a limited group of people – hundreds, or maybe thousands. Then it was introduced to the public and it was wrong . . . to introduce it in the same way." See Vivian Yeo, "Microsoft OneCare was 'Good Enough'," *ZDNet*, October 16, 2009, http://www.zdnet.com/microsoft-onecare-was-good-enough-2062058697.

172 **The former U.S. counterterrorism czar:** Richard Clarke warns about the growing cyber-war threat in Richard A. Clarke and Robert K. Knake, *Cyber War: The Next Threat to National Security and What to Do About It* (New York: HarperCollins, 2010). Researchers warn against the alarmist rhetoric about cyber threats and the emergence of a cyber-industrial complex in the United States in Jerry Brito and Tate Watkins, *Loving the Cyber Bomb? The Dangers of Threat Inflation in Cybersecurity Policy*, Working Paper no. 11–24, Washington: George Mason University, 2011. David Perera traces the history of the term "Electronic Pearl Harbor" from its first public usage in 1996 to the present in "Stop Saying 'Cyber Pearl Harbor,'" *FierceGovernmentIT*, June 13, 2012, http://www.fiercegovernmentit.com/story/stop-saying-cyber-pearl-harbor/2012-06-13. Thomas Rid argues that cyber war is unlikely to occur in the future in "Cyber War Will Not Take Place," *Journal of Strategic Studies* 35, no. 1 (2012).

173 **Kaspersky was back in the news:** Kaspersky Lab's announcement on the discovery of Flame in "Kaspersky Lab and ITU Research Reveals New Advanced Cyber Threat," is at Kaspersky Lab, May 28, 2012, http://www.kaspersky.com/about/news/virus/2012/Kaspersky_Lab_and_

ITU_Research_Reveals_New_Advanced_Cyber_Threat. For discussion, see Chris Bronk, "Cyber Intrigue: The Flame Malware International Politics," *Cyber Dialogue*, May 31, 2012, http://www.cyberdialogue. ca/2012/05/cyber-intrigue-the-flame-malware-international-politics; and Tom Gjelten, "'Flame' Virus Fuels Political Heat Over Cyber Threats," KQED News, June 2, 2012, http://www.kqed.org/news/ story/2012/06/02/96069/flame_virus_fuels_political_heat_over_cyber_ threats?source=npr&category=technology.

173 **the ITU is the world's oldest international organization:** Milton Mueller has written on the politics of international Internet governance in *Networks and States: The Global Politics of Internet Governance* (Cambridge: The MIT Press, 2010).

174 **proposed a "code of conduct":** In 2011, Russia, China, Tajikistan, and Uzbekistan proposed a voluntary code of conduct for cyberspace at the United Nations. See letter dated 12 September 2011 from the Permanent Representatives of China, the Russian Federation, Tajikistan and Uzbekistan to the United Nations addressed to the Secretary-General, available at: http://www.cs.brown.edu/courses/csci1800/ sources/2012_UN_Russia_and_China_Code_o_Conduct.pdf; and Nate Anderson, "Russia, China, Tajikistan propose UN 'code of conduct' for the 'Net," *Ars Technica*, September 20, 2011, http://arstechnica.com/ tech-policy/2011/09/russia-china-tajikistan-propose-un-code-of-conduct-for-the-net.

174 **connections between Flame and another devastating cyber weapon, Stuxnet:** The Kaspersky Flame FAQ is available at: "The Flame: Questions and Answers," *Secure List*, May 28, 2012, http://www. securelist.com/en/blog/208193522. The connection between Flame and Stuxnet is discussed in Jim Finkle and Joseph Menn, "Some Flame Code Found in Stuxnet Virus: Experts," Reuters, June 12, 2012, http://www. reuters.com/article/2012/06/12/us-media-tech-summit-flame-idUS-BRE85A0TN20120612; Greg Miller, Ellen Nakashima, and Julie Tate, "U.S., Israel Developed Flame Computer Virus to Slow Iranian Nuclear Efforts, Officials Say," *Wall Street Journal*, June 19, 2011, http://www. washingtonpost.com/world/national-security/us-israel-developed-computer-virus-to-slow-iranian-nuclear-efforts-officials-say/2012/06/19/ gJQA6xBPoV_story.html; and Kenneth Rapoza, "Kaspersky Lab: Same Countries Behind Stuxnet and Flame Malware," *Forbes*, June 11, 2012,

http://www.forbes.com/sites/kenrapoza/2012/06/11/kaspersky-lab-same-countries-behind-stuxnet-and-flame-malware/.

11: STUXNET AND THE ARGUMENT FOR CLEAN WAR

176 **a detailed "decoding" of the virus:** For Langner's research on Stuxnet, visit his blog at http://www.langner.com/en/blog/. See also Ralph Langner, "Stuxnet: Dissecting a Cyberwarfare Weapon," *Security & Privacy*, IEEE 9, no.3 (2011): 49–51.

177 **the planning and operational process behind the Stuxnet virus:** On June 1, 2012, the *New York Times* reported that anonymous current and former government officials of the U.S., Europe, and Israel had confirmed that Stuxnet was indeed the work of American and Israeli experts, under orders of President Obama, who wanted to slow Iran's progress towards building an atomic bomb without launching a traditional attack. See David Sanger, "Obama Order Sped Up Wave of Cyberattacks Against Iran," http://www.nytimes.com/2012/06/01/world/middleeast/obama-ordered-wave-of-cyberattacks-against-iran.html?pagewanted=all&_r=0. Sanger's article was adapted from his book, *Confront and Conceal: Obama's Secret Wars and Surprising Use of American Power* (New York: Crown Publishers, 2012). See also William J. Broad, John Markoff and David E. Sanger, "Israeli Test on Worm Called Crucial in Iran Nuclear Delay," *New York Times*, January 15, 2011, http://www.nytimes.com/2011/01/16/world/middleeast/16stuxnet.html?pagewanted=2&_r=1&hp; and William J. Broad and David E. Sanger, "Worm Was Perfect for Sabotaging Centrifuges," *New York Times*, November 18, 2010, http://www.nytimes.com/2010/11/19/world/middleeast/19stuxnet.html.

177 **the kinds of manoeuvres that could exploit holes:** The Siemens and Idaho National Lab 2008 presentation of the PCS7's vulnerabilities to cyber attacks is available at Marty Edwards and Todd Stauffer, "Control System Security Assessments," Presentation prepared for the 2008 Siemens Automation Summit, http://graphics8.nytimes.com/packages/pdf/science/NSTB.pdf.

178 **code behind Stuxnet was far larger than a typical worm:** Symantec reversed engineered Stuxnet and documented its findings in Nicolas Falliere, Liam Ó Murchú, and Eric Chien, "W32.Stuxnet

Dossier Version 1.4," *Symantec*, February 2011, http://www.symantec. com/content/en/us/enterprise/media/security_response/whitepapers/ w32_stuxnet_dossier.pdf.

178 **an obscure date in the worm's code:** The clues of Israeli involvement in Stuxnet's code have been reported by Michael Joseph Gross in "A Declaration of Cyberwar," *Vanity Fair*, April 2011, http://www. vanityfair.com/culture/features/2011/04/stuxnet-201104, 4; Paul Roberts, "Stuxnet Analysis Supports Iran-Israel Connections," *Threat Post*, September 30, 2010, http://threatpost.com/en_us/blogs/stuxnet-analysis-supports-iran-israel-connections-093010; John Markoff and David E. Sanger, "In a Computer Worm, a Possible Biblical Clue," *New York Times*, September 29, 2010, http://www.nytimes.com/2010/09/30/ world/middleeast/30worm.html?pagewanted=all&_r=0; and William J. Broad and David E. Sanger, "Worm Was Perfect for Sabotaging Centrifuges," *New York Times*, November 18, 2010, http://www.nytimes. com/2010/11/19/world/middleeast/19stuxnet.html.

178 **an Iranian double agent working for Israel:** Richard Sale reported on how Iranian control systems were infected by Stuxnet in, "Stuxnet Loaded by Iran Double Agents," *Industrial Safety and Security Source*, April 11, 2012, http://www.isssource.com/stuxnet-loaded-by-iran-double-agents. See also Dorothy E. Denning, "Stuxnet: What Has Changed," *Future Internet* 4, no.3 (2012): 672–687.

181 **high-tech means of fighting clean wars:** James Der Derian writes about "virtuous war" in *Virtuous War: Mapping the Military-Industrial-Media-Entertainment Network* (New York: Routledge, 2009). See also Jennifer Leonard, "James Der Derian on Imagining Peace," *Renegade Media*, http://www.renegademedia.info/books/james-derderian.html.

182 **Writing in the *Bulletin of the Atomic Scientists*:** R. Scott Kemp analyzes the implications of developing offensive cyber capabilities in "Cyberweapons: Bold Steps in a Digital Darkness?," *Bulletin of the Atomic Scientists*, June 7, 2012, http://www.thebulletin.org/web-edition/op-eds/ cyberweapons-bold-steps-digital-darkness.

183 **thirty-three states included cyber warfare in their military planning:** James A. Lewis and Katrina Timlin review the policies and organizations of 133 states to determine how they are organized to deal

with cyber security in "Cybersecurity and Cyberwarfare," Center for Strategic and International Studies, 2011; available at: http://www.unidir.org/pdf/ouvrages/pdf-1-92-9045-011-J-en.pdf.

183 **Some, like India, boast about developing offensive cyber attack capabilities:** On June 11, 2012, the *Times of India* reported on India's National Security Council's plan to allow the Defence Intelligence Agency and National Technical Research Organization to carry out cyber offensives against other countries if necessary, in Josy Joseph, "India to Add Muscle to its Cyber Arsenal," http://articles.timesofindia.indiatimes.com/2012-06-11/india/32174336_1_cyber-attacks-offensive-cyber-government-networks.

184 **1,800 cases of fake electronic components:** The case of counterfeit chips in the flight computer of an F-15 fighter jet at Robins Air Force Base was reported on by Brian Burnsed, Cliff Edwards, Brian Grow, and Chi-Chu Tschang, in "Dangerous Fakes," *Business Week*, October 2, 2008, http://www.businessweek.com/magazine/content/08_41/b4103034193886.htm.

185 **via the SHODAN search tool anyone could discover MAC addresses:** According to its website, the SHODAN search engine (developed by John Matherly) is "a search engine that lets you find specific computers (routers, servers, etc.) using a variety of filters. Some have also described it as a public port scan directory or a search engine of banners." In "Cyber Search Engine Shodan Exposes Industrial Control Systems to New Risks," *Washington Post*, June 3, 2012, http://www.washingtonpost.com/investigations/cyber-search-engine-exposes-vulnerabilities/2012/06/03/gJQAIK9KCV_story.html, journalist Robert O'Harrow Jr. wrote: "Matherly and other Shodan users quickly realized they were revealing an astonishing fact: Uncounted numbers of industrial control computers, the systems that automate such things as water plants and power grids, were linked in, and in some cases they were wide open to exploitation by even moderately talented hackers."

185 **"I was walking down the street . . .":** Kim Zetter reported on the RuggedCom vulnerability in "Equipment Maker Caught Installing Backdoor Account in Control System Code," *Wired*, April 25, 2012, http://www.wired.com/threatlevel/2012/04/ruggedcom-backdoor.

186 **we are building a digital edifice for the entire planet, which sits above us like a house of cards:** Supply-chain vulnerabilities have been documented in Committee on Armed Services United States Senate, "Inquiry into Counterfeit Electronic Parts in the Department of Defence Supply Chain, Report 112-167," May 21, 2012, available at: http://armed-services.senate.gov/Publications/Counterfeit%20 Electronic%20Parts.pdf; Marcus H. Sachs, "Can We Secure the Information Technology Supply Chain in the Age of Globalization?" Verizon, http://crissp.poly.edu/media/sachs_slides.pdf; and Dana Gardner, "Corporate Data, Supply Chains Remain Vulnerable to Cyber Crime Attacks, says Open Group Conference Speaker," *ZDNet*, June 5, 2012, http://www.zdnet.com/blog/gardner/corporate-data-supply-chains-remain-vulnerable-to-cyber-crime-attacks-says-open-group-conference-speaker/4644.

The Russian vendor Positive Technologies found alarming statistics about SCADA system vulnerabilities based on an analysis of vulnerabilities in databases like ICS-CERT, Siemens' Product CERT, exploit-db, and vendor advisories. They found "the number of security flaws found within ten months is far bigger than the number of flaws found during the whole previous period starting from 2005." Positive Technologies documented its findings in Yury Goltsev et al., SCADA Safety in Numbers, Positive Technologies, 2012, available at: www.ptsecurity.com/download/SCADA_analytics_english.pdf.

186 **"Cyberwar is very different from nuclear war . . .":** Fred Kaplan, "Why the United States Can't Win a Cyberwar," *Slate*, June 8, 2012, http://www.slate.com/articles/news_and_politics/war_stories/2012/06/obama_s_cyber_attacks_on_iran_were_carefully_considered_but_the_nuclear_arms_race_offers_important_lessons_.html.

12: THE INTERNET IS OFFICIALLY DEAD

188 **The June 2011 RSA breach hit the American security:** "Breachfest 2011" is documented in Matt Liebowitz, "2011 Set to Be Worst Year Ever for Security Breaches," *Tech News Daily*, June 10, 2011, http://www.technewsdaily.com/2710-2011-worst-year-ever-security-breaches.html.

191 **I first read about Narus's technology:** Narus's 2007 press release is available at "Narus Expands Traffic Intelligence Solution to Webmail Targeting," *Narus*, December 10, 2007, http://www.narus.com/index.

php/overview/narus-press-releases/press-releases-2007/274-narus-expands-traffic-intelligence-solution-to-webmail-targeting.

192 **its sales to Telecom Egypt:** Timothy Karr discusses the use of Narus in Egypt in "One U.S. Corporation's Role in Egypt's Brutal Crackdown," *Huffington Post,* January 28, 2011, http://www.huffington-post.com/timothy-karr/one-us-corporations-role-_b_815281.html.

192 **After thirty-three years of active service:** In *The Shock Doctrine,* Naomi Klein argues that Kenneth Minihan is responsible for implementing the "disaster capitalism complex," defined as "a fully fledged new economy in homeland security, privatised war and disaster reconstruction tasked with nothing less than building and running a privatised security state, both at home and abroad." Similarly, in his book *Spies for Hire,* investigative journalist Tim Shorrock traces the subservience of public to private interests in the intelligence-contracting industry, an industry that specifically "serves the needs of government and its intelligence apparatus." Shorrock writes, "In the past, Minihan said, contractors 'used to support military operations; now we participate [in them]. We're inextricably tied to the success of their operations.'" Naomi Klein, *The Shock Doctrine: The Rise of Disaster Capitalism* (New York: Henry Holt and Company, 2007); and Tim Shorrock, *Spies for Hire: The Secret World of Intelligence Outsourcing* (New York: Simon & Schuster, 2008).

13: A ZERO DAY NO MORE

195 **In the aftermath of the 2011 revolution:** The chaos that followed the collapse of regimes in Egypt and Libya helped pry open secretive security apparatuses, revealing the extent of their international linkages. See Steve Ragan, "Report: U.K. Firm Offered IT Intrusion Tools to Egyptian Government," *Tech Herald,* April 27, 2011, http://www.thetechherald.com/articles/Report-U-K-firm-offered-IT-intrusion-tools-to-Egyptian-government; Karen McVeigh, "British Firm Offered Spying Software to Egyptian Regime – Documents," *Guardian,* April 28, 2011, http://www.guardian.co.uk/technology/2011/apr/28/egypt-spying-software-gamma-finfisher; Matt Bradley, Paul Sonne, and Steve Stecklow, "Mideast Uses Western Tools to Battle the Skype Rebellion," *Wall Street Journal,* June 1, 2011, http://online.wsj.com/article/SB1000142405270230 45208045763459708624200038.html; and Mikko Hyppönen, "Egypt, FinFisher Intrusion Tools and Ethics," *F-Secure,* March 8, 2011, https://

www.f-secure.com/weblog/archives/00002114.html. See also John
Scott-Railton, *Revolutionary Risks: Cyber Technology and Threats in the 2011
Libyan Revolution*, CIWAG Case Studies Series, forthcoming, 2013.

196 **Among the brochures in the "Spy Files":** The "Spy Files" can be
accessed at "The Spy Files," *WikiLeaks*, http://wikileaks.org/the-spyfiles.
html. See Ronald Deibert, "Big Data Meets Big Brother," *Privacy
International*, November 30, 2011, https://www.privacyinternational.org/
opinion-pieces/big-data-meets-big-brother.

200 **a glimpse into a vast labyrinth and arms race in cyberspace:**
The Citizen Lab, led by Morgan Marquis-Boire, has found a growing
commercial market for offensive computer network intrusion capabilities developed by companies in Western democratic countries. See
"Backdoors are Forever: Hacking Team and the Targeting of Dissent?,"
October 10, 2010, https://citizenlab.org/2012/10/backdoors-are-forever-
hacking-team-and-the-targeting-of-dissent/; "The SmartPhone Who
Loved Me: FinFisher Goes Mobile?," August 29, 2012, https://citizenlab.
org/2012/08/the-smartphone-who-loved-me-finfisher-goes-mobile/;
and "From Bahrain With Love: FinFisher's Spy Kit Exposed?," July 25,
2012, https://citizenlab.org/2012/07/from-bahrain-with-love-finfishers-
spy-kit-exposed/. FinFisher's FinSpy brochure is available at FinSpy:
"Remote Monitoring and Infection Solutions," http://wikileaks.org/
spyfiles/docs/gamma/289_remote-monitoring-and-infection-solutions-
finspy.html.

200 **"The cyber domain of computers and related electronic
activities . . .":** Nye describes the characteristics of cyberspace that
lend the domain to arms racing in Joseph S. Nye, "Cyber War and
Peace," Al-Jazeera, April 21, 2012, http://www.aljazeera.com/indepth/
opinion/2012/04/201241510242769575.html.

204 **In 2011, the German hacker collective, Chaos Computer Club:**
The Chaos Computer Club's discovery of the "State Trojan" has been
documented in "Chaos Computer Club Analyzes Government
Malware," *Chaos Computer Club*, October 8, 2010, http://www.ccc.de/
en/updates/2011/staatstrojaner; Elinor Mills, "Trojan Opened Door to
Skype Spying," CBS News, October 10, 2011, http://www.cbsnews.
com/2100-205_162-20118260.html; Bob Sullivan, "German Officials
Admit Using Spyware on Citizens, As Big Brother Scandal Grows,"

NBC News, October 11, 2011, http://redtape.nbcnews.com/_news/ 2011/10/11/8274668-german-officials-admit-using-spyware-on-citizens- as-big-brother-scandal-grows?lite; and Bob Sullivan, "Chaos Computer Club: German Gov't Software Can Spy on Citizens," NBC News, October 8, 2011, http://redtape.nbcnews.com/_news/2011/ 10/08/8228095-chaos-computer-club-german-govt-software-can-spy- on-citizens?lite.

204 **a Bangkok middleman:** Andy Greenberg profiled "The Grugq" and the exploits market in "Shopping For Zero-Days: A Price List For Hackers' Secret Software Exploits," *Forbes*, March 23, 2012, http://www. forbes.com/sites/andygreenberg/2012/03/23/shopping-for-zero-days- an-price-list-for-hackers-secret-software-exploits/.

206 **One of the few companies not afraid to speak out:** For more information on VUPEN, see Andy Greenberg, "Meet the Hackers Who Sell Spies the Tools to Crack Your PC (And Get Paid Six-Figure Fees)," *Forbes*, March 21, http://www.forbes.com/sites/andygreenberg/2012/03/21/ meet-the-hackers-who-sell-spies-thetools-to-crack-your-pc-and-get- paid-six-figure-fees/. See also Greenberg's, "New Grad Looking For a Job? Pentagon Contractors Post Openings For Black-Hat Hackers," *Forbes*, June 15, 2012, http://www.forbes.com/sites/andygreenberg/2012/06/15/ new-grad-looking-for-a-job-pentagon-contractors-post-openings-for- black-hat-hackers-2.

207 **a service offered by one U.S. company, Endgame:** Endgame is extensively profiled in Michael Riley and Ashlee Vance, "Cyber Weapons: The New Arms Race," *Business Week*, July 20, 2011, http:// www.businessweek.com/magazine/cyber-weapons-the-new-arms- race-07212011.html.

209 **Hacking Team:** The use of Hacking Team products is detailed in Vernon Silver, "Spyware Leaves Trail to Beaten Activist Through Microsoft Flaw," *Bloomberg News*, October 10, 2012, http://www. bloomberg.com/news/2012-10-10/spyware-leaves-trail-to-beaten- activist-through-microsoft-flaw.html; and Nicole Perlroth, "Ahead of Spyware Conference, More Evidence of Abuse," *New York Times*, October 10, 2012, http://bits.blogs.nytimes.com/2012/10/10/ahead-of- spyware-conference-more-evidence-of-abuse/.

211 **the NSA partners with "cleared" universities to train students:**
The phenomenon of cyber-ops courses in universities in the United
States is profiled in "Exclusive: Spy Agency Seeks Cyber-ops
Curriculum," Reuters, May 22, 2012, http://ca.reuters.com/article/
technologyNews/idCABRE84L12T20120522?pageNumber=1&virtualB
randChannel=0.

211 **Privacy International has identified at least thirty British
companies:** See Jamie Doward and Rebecca Lewis, "UK Exporting
Surveillance Technology to Repressive Nations," *Guardian*, April 7, 2012,
http://www.guardian.co.uk/world/2012/apr/07/surveillance-technol-
ogy-repressive-regimes.

212 **In August 2011 a French company, Amesys:** See Margaret Coker
and Paul Sonne, "Firms Aided Libyan Spies," *Wall Street Journal*, August
30, 2011, http://online.wsj.com/article/SB10001424053111904199404576
538721260166388.html.

212 **In July 2011, the *Washington Post* reported on a U.S. Air Force
contract solicitation:** Detailed in Walter Pincus, "U.S. Plans to Provide
Iraq with Wiretapping System," *Washington Post*, July 30, 2011, http://
www.washingtonpost.com/world/national-security/us-plans-to-pro-
vide-iraq-with-wiretapping-system/2011/07/26/gIQAGexvjI_story.html.

212 **Swedish television producers uncovered a huge surveillance
market:** In May 2012, the Swedish news show *Uppdrag Granskning*
uncovered the links between TeliaSonera and Central Asian governments.
See Eva Galperin, "Swedish Telcom Giant Teliasonera Caught Helping
Authoritarian Regimes Spy on Their Citizens," Electronic Frontier
Foundation, May 18, 2012, https://www.eff.org/deeplinks/2012/05/
swedish-telcom-giant-teliasonera-caught-helping-authoritarian-
regimes-spy-its.

213 **Bloomberg concluded that the technology:** Ben Elgin, Alan Katz,
and Vernon Silver reported that Ericsson, Creativity Software, and
AdaptiveMobile had been providing surveillance equipment to the
government of Iran in, "Iranian Police Seizing Dissidents Get Aid of
Western Companies," Bloomberg News, October 30, 2011, http://www.
bloomberg.com/news/2011-10-31/iranian-police-seizing-dissidents-get-
aid-of-western-companies.html.

213 **Nokia Siemens Networks faced an international:** In August 2011, it was reported that Bahraini dissidents arrested by authorities were presented with transcripts of their own text messages during interrogations, and the capacity to intercept the text messages was acquired through equipment from Nokia Siemens Networks, based in Finland, and trovicor, a German company. See Ben Elgin and Vernon Silver, "Torture In Bahrain Becomes Routine With Help From Nokia Siemens," Bloomberg News, August 22, 2011, http://www.bloomberg.com/news/2011-08-22/torture-in-bahrain-becomes-routine-with-help-from-nokia-siemens-networking.html.

215 **". . . information technology, unlike bombs or tanks, is fundamentally multi-purpose in nature . . .":** The issue of how to control the digital arms trade is contentious. For differing views, see Milton Mueller, "Technology As Symbol: Is Resistance to Surveillance Technology Being Misdirected?" Internet Governance Project, December 20, 2011, http://www.internetgovernance.org/2011/12/20/technology-as-symbol-is-resistance-to-surveillance-technology-being-misdirected; and Member of the European Parliament Marietje Schaake's proposal, detailed in "European Parliament Endorses Stricter European Export Control of Digital Arms," October 23, 2012, http://www.marietjeschaake.eu/2012/10/ep-steunt-d66-initiatief-controle-europese-export-digitale-wapens. In November 2012, the United States Department of State issued a guidance document that attempted to clarify under what conditions companies might violate restrictions on the export of "sensitive technologies" to countries like Iran and Syria, which can be found at: https://www.federalregister.gov/articles/2012/11/13/2012-27642/department-of-state-state-department-sanctions-information-and-guidance#h-10. See also Ben Wagner, *Exporting Censorship and Surveillance Technology* (The Hague: Hivos, 2012).

14: ANONYMOUS: EXPECT US

217 **Epigraph:** Lewis Mumford, *The Pentagon of Power: The Myth of the Machine, Vol. II* (New York: Harcourt Brace Jovanovich, 1974). Mumford's *Pentagon of Power* is a major influence on my thinking about political resistance and technology. I assigned it as the standard text to my graduate seminar on the Politics of Planetary Surveillance, taught at the University of Toronto from 1997 to 2004.

221 **Ryan Cleary, a nineteen-year-old member:** Details of Cleary's arrest appear in Graham Cluley, "Ryan Cleary has Asperger's Syndrome, Court Hears," *Sophos Naked Security*, June 26, 2011, http://nakedsecurity. sophos.com/2011/06/26/ryan-cleary-aspergers-syndrome.

222 **Anonymous's breaches are typically followed by the exfiltration of data:** For details on the Stratfor breach, see Richard Norton-Taylor and Ed Pilkington, "Hackers Expose Defence and Intelligence Officials in US and UK," *Guardian*, January 8, 2010, http://www.guardian.co.uk/technology/2012/jan/08/hackers-expose-defence-intelligence-officials.

223 **Neustar . . . surveyed IT professionals:** Neustar reports on the impacts of DDOS attacks in *Neustar Insights*, "DDOS Survey: Q1 2012 When Businesses Go Dark," http://hello.neustar.biz/rs/neustarinc/images/neustar-insights-ddos-attack-survey-q1-2012.pdf.

223 **The New York–based hacker and artist collective:** Details of Electronic Disturbance Theatre's use of DDOS attacks in support of the Zapatista movement are available in Coco Fusoco, "Performance Art in a Digital Age: A Conversation with Ricardo Dominguez," *The Hacktivist Magazine* (2001), http://www.iwar.org.uk/hackers/resources/the-hacktivist/issue-1/vol1.html.

224 **has likened them to picket lines:** Evgeny Morozov argues that "under certain conditions . . . DDOS attacks can be seen as a legitimate expression of dissent, very much similar to civil disobedience" in "In Defense of DDOS," *Slate*, December 13, 2010, http://www.slate.com/articles/technology/technology/2010/12/in_defense_of_ddos.html. The potential of hacktivism as an agent of political change is discussed in Mark Manion and Abby Goodrum, "Terrorism or Civil Disobedience: Toward a Hacktivist Ethic," in *Internet Security: Hacking, Counterhacking, and Society*, ed. Kenneth Einar Himma (Sudbury: Jones and Bartlett Publishers, 2007). After the success of Operation: Tunisia, Reporters Without Borders reported that the solidarity action had the unintended consequence of the arrest of Tunisian bloggers and online activists; see "Wave of Arrests of Bloggers and Activists," *Reporters Without Borders*, January 7, 2011, http://en.rsf.org/tunisia-wave-of-arrests-of-bloggers-and-07-01-2011,39238.html. Steven Murdoch writes about the "double-edged sword" of digital activism in "Destructive Activism: The Double-Edged Sword of Digital Tactics,"

in *Digital Activism Decoded* ed. Mary Joyce, (New York: iDebate Press, 2010), 137–148.

224 **Anonymous's Operation Tunisia:** Details of Anonymous's launching of DDOS attacks on eight Tunisian government websites during the 2011 Tunisian uprisings are available in Yasmine Ryan, "Tunisia's Bitter Cyberwar," Al-Jazeera, January 6, 2011, http://www.aljazeera.com/indepth/features/2011/01/201116141458393962.html. Anonymous attacks during the 2011 Egyptian uprisings are detailed in Paul Wagenseil, "Anonymous 'Hacktivists' Attack Egyptian Websites," NBC News, January 26, 2011, http://www.msnbc.msn.com/id/41280813/ns/technology_and_science-security/t/anonymous-hacktivists-attack-egyptian-websites/. In December 2010, Anonymous launched a DDOS protest against the website of the Zimbabwe African National Union – Patriotic Front (ZANU-PF); see "Operation Zimbabwe Success," *AnonNews*, December 29, 2010, http://anonnews.org/press/item/94/. In June 2011, Anonymous launched attacks on ninety-one websites, including fifty-one Malaysian government sites; see Niluksi Koswanage and Liau Y-Sing, "Hackers Disrupt 51 Malaysian Government Websites," Reuters, June 16, 2011, http://www.reuters.com/article/2011/06/16/us-malaysia-hackers-idUS-TRE75F06Y20110616. The Anonymous movement was split on the Libyan uprisings; see "Operation Reasonable Reaction," *Github*, https://github.com/bibanon/bibanon/wiki/Operation-Reasonable-Reaction. The relationship between the Occupy Movement and Anonymous is detailed in Sean Captain, "The Real Role of Anonymous in Occupy Wall Street," *Fast Company*, October 17, 2011, http://www.fastcompany.com/1788397/the-real-role-of-anonymous-at-occupy-wall-street.

225 **is it wise to actually encourage DDoS attacks:** Yochai Benkler explains why Anonymous should not be viewed as a threat to national security in "Hacks of Valor," *Foreign Affairs*, April 4, 2012, http://www.foreignaffairs.com/articles/137382/yochai-benkler/hacks-of-valor.

225 **One of the few to study this question in depth:** Gabriella Coleman's work offers a comprehensive history and analysis of Anonymous: Gabriella Coleman "Our Weirdness Is Free: The Logic of Anonymous – Online Army, Agent Chaos, and Seeker of Justice," *Triple Canopy* (2012), http://canopycanopycanopy.com/15/our_weirdness_is_free; and "Peeking Behind the Curtain at Anonymous: Gabriella Coleman at TEDGlobal 2012," TED Blog, June 27, 2012, http://blog.ted.

com/2012/06/27/peeking-behind-the-curtain-at-anonymous-gabriella-coleman-at-tedglobal-2012/.

228 **MIT Museum Hack archivist:** A history of MIT hacks is detailed in T.F. Peterson, *Nightwork: A History of Hacks and Pranks at MIT*, (Cambridge: MIT Press, 2011). See especially Brian Leibowitz, "A Short History of the Terminology," in *Nightwork*, ed. T.F. Peterson. See also Steven Levy, *Hackers: Heroes of the Computer Revolution* (Sebastopol: O'Reilly Media, Inc., 2010).

228 **". . . and the one used in the media":** Molly Sauter's SXSW presentation on media portrayals of hackers is available at "Policy Effects of Media Portrayals of Hacktivists," SXSW, http://schedule.sxsw. com/2012/events/event_IAP12520.

230 **numerous examples of security research being stifled:** The Electronic Frontier Foundation traces the chilling effects of the "anti-circumvention" provisions in the DMCA on research in "Unintended Consequences: Twelve Years Under the DMCA," https://www.eff.org/wp/unintended-consequences-under-dmca/#footnoteref13_pdu4ggq. See also Marcia Hofmann and Katitza Rodriguez, "Coders' Rights at Risk in the European Parliament," Electronic Frontier Foundation, June 20, 2012, https://www.eff.org/deeplinks/2012/06/eff-european-parliament-directive-attack-information-systems.

15: TOWARDS DISTRIBUTED SECURITY AND STEWARDSHIP IN CYBERSPACE

232 **Epigraph:** H.G. Wells (1866–1946) was one of the most prolific authors in the English language. Although most well known for his works of science fiction, he also published many essays and books on history and politics. He was a proponent of world government along republican lines ("The New Republic"), and supported the League of Nations. He was also the principal author of the *Sankey Declaration of the Rights of Man* (1940), which was later superseded by the Universal Declaration of Human Rights. H.G. Wells, *Anticipations of the Reaction of Mechanical and Scientific Progress Upon Human Life and Thought* (New York: Harper & Brothers, 1901).

232 **"our central nervous system in a global embrace":** The quotation is from Marshall McLuhan, *Understanding Media*, page 3: "Today, after

more than a century of electric technology, we have extended our central nervous system in a global embrace, abolishing both space and time as far as our planet is concerned."

234   **"Internet Freedom in a Suitcase":** The phrase refers to the so-called Internet-in-a-suitcase project that was developed by the New America Foundation's Open Technology Initiative, and funded by the U.S. Department of State. See James Glanz and John Markoff, "US Underwrites Internet Detour Around Censors," *New York Times,* June 12, 2011 http://www.nytimes.com/2011/06/12/world/12internet. html?pagewanted=all&_r=0.

234   **Faced with mounting problems and pressures:** I developed the ideas in this chapter in prior articles and essays, including: "Protect the Net: The Looming Destruction of the Global Communications Environment," in Mark Kingwell and Patrick Tunnel, eds., *Rites of Way: The Politics and Poetics of Public Space* (Waterloo: Wilfred Laurier Press, 2009); "Towards a Cyber Security Strategy for Global Civil Society?," *Global Information Society Watch* (South Africa: APC and Hivos, 2011); and "The Growing Dark Side of Cyberspace (. . . and What To Do About It)," *Penn State Journal of Law & International Affairs* 1, iss. 2 (2012). I first wrote about the importance of civil society taking seriously Internet design issues in "The Politics of Internet Design: Securing the Foundations for Global Civil Society Networks," in ed. Stephen Coleman, *The E-Connected World: Its Social and Political Implications* (McGill University Press, 2003) and in "Deep Probe: The Evolution of Network Intelligence," *Intelligence and National Security*, Volume 17, No 1, 2004.

238   **"negarchy" as a structural alternative:** For more on distributed security, see Daniel Deudney, *Bounding Power: Republican Security Theory from the Polis to the Global Village* (Princeton: Princeton University Press, 2007).

242   **The first custodians of the Internet believed that it did:** See http://www.cyberdialogue.ca/previous-dialogues/2012-about/papers/ for a collection of essays on the topic of "Stewardship in Cyberspace," commissioned for the University of Toronto's second annual Cyber Dialogue, March 2012.

244 **Universities have a special role to play:** For a discussion of the evolution of the university education system from an elite to popular institution, see Peter Scott, *The Meanings of Mass Higher Education* (Buckingham, UK: Open University Press, 1995).

My inspiration for the role of the university in securing an open global commons of communication and information is primarily due to the ideas on the matter articulated by John Dewey. Dewey outlined the central importance of communications to democracy, and some of the problems that plague a viable communications sphere for the public, in *The Public and its Problems: An Essay in Political Inquiry* (Chicago: Gateway Books, 1946).

ACKNOWLEDGEMENTS

I have often said that the Citizen Lab is unique by nature of our worldwide collaborative partnerships. For making this book possible, I am extremely grateful to the following exceptionally talented Citizen Lab staff, fellows, associates, visiting fellows, et cetera, past and present, for their support, initiative, and ingenuity: Graeme Bunton, Michelle Levesque, Michael Hull, Jeremy Vernon, Jonathan Doda, Clayton Epp, Elias Adum, Eran Henig, François Fortier, Matthew Carrieri, Jakub Dalek, Robert Guerra, Seth Hardy, Saad Khan, Katie Kleemola, Sarah McKune, Helmi Noman, Irene Poetranto, Lidija Sabados, Adam Senft, Igor Valentovitch, Greg Wiseman, Ali Bangi and all of ASL19, Cherise Seucharan, Jennie Phillips, James Tay, Karl Kathuria, Brenden Kuerbis, Phillipa Gill, Eneken Tikk-Ringas, Christopher Bronk, Rex Hughes, Roger Hurwitz, Camino Kavanagh, Luis Horacio Najera, Jon Penney, John Sheldon, Chris Davis, Byron Sonne, Peter Wills, Morgan Marquis-Boire, John Scott-Railton, Nart Villeneuve, and Greg Walton. Special thanks to the Citizen Lab research manager, Masashi Crete-Nishihata.

The OpenNet Initiative (ONI) has been a ten-year partnership between the Citizen Lab and the Berkman Center for Internet & Society at Harvard University, and, for a period of time, the Oxford Internet Institute, and the University of Cambridge (later, the SecDev Group). Included in ONI's extended community are regional research networks, such as OpenNet.Eurasia, OpenNet.

Asia, and the Cyber Stewards project. Thanks to all of my colleagues in these networks, and in particular to Rafal Rohozinski, John Palfrey, Jonathan Zittrain, Robert Faris, Jillian York, Rebekah Heacock, and Tattugul Mambetalieva.

The Koobface, GhostNet, and Shadows reports were undertaken under the auspices of the Information Warfare Monitor, a collaboration between the Citizen Lab and the SecDev Group, from 2002–2011. Thanks to Steven Adair of the Shadowserver Foundation, who collaborated with us on *Shadows in the Cloud*.

Special thanks to the Tibetan community worldwide, especially Lhadon Tethong, Students for a Free Tibet, the Tibetan Government-in-Exile, and the Office of His Holiness the Dalai Lama.

Research funding for the Citizen Lab has come from the generous support of the Canada Centre for Global Security Studies, the John D. and Catherine T. MacArthur Foundation, IDRC, and Google Inc. Special thanks to Eric Sears, Laurent Elder, Matthew Smith, and Bob Boorstin.

The Citizen Lab's success is, in part, attributable to the strong base of support we have enjoyed at the University of Toronto's Munk School of Global Affairs and the Department of Political Science. Thanks in particular to my colleagues Janice Stein, Lou Pauly, David Cameron, and President David Naylor for their support, and to Franklyn Griffiths for suggesting that I create a lab in the first place. I am very grateful to the Munk School staff, especially Margaret McKone, Wilhelmina Peter, Miriam McTiernan, Vanessa Johan, Lucinda Li, and Sean Willett.

I am particularly grateful to Vanessa van den Boogaard and Marianne Lau for outstanding research assistance.

Thanks to Doug Pepper and Jenny Bradshaw at McClelland & Stewart, and to Michael Levine.

I am exceptionally grateful to my editor, Ken Alexander, from whom I have learned immeasurably over the process of writing this book. In addition to being an exceptional editor, Ken has been a cheerleader, coach, and friend. His tireless attention to detail and enthusiasm about the craft of writing are truly inspiring. I benefited directly from his many careful interventions and thoughtful feedback, as well as his informal guidance and advice.

Special thanks to Mark Zacher at the University of British Columbia, who showed me the value of working collaboratively, of paying constant attention to principles, and of appreciating the responsibilities that come with being a professional scholar.

Thank you to my children – Emily, Rosalind, Ethan, and Michael – for showing pride and interest in all things related to the Citizen Lab, and for caring about making the world around you a better place.

Finally, I would like to thank and acknowledge my wife, Jane Gowan, who has supported me throughout not only this book project, but through all of my professional activities. By the example she sets with her own art and work, I've learned from her what loving dedication to a craft really means. Thank you, Janie.

INDEX

The letter "n" following a page number indicates that the information is in a note.

against Burmese opposition, 166
by Anonymous, 146, 155, 219-20, 222
by SEA, 155, 160
criminalization of, 228
Cult of the Dead Cow, 224
of Dalai Lama's computers, 21-22, 161
of Google networks, 61, 115
of U.S. military and intelligence agencies, 23
origins, 227-28
hacktivists and hacktivism, defined, 226-27
Hainan Island, 24-25, 26
Haiti, beneficial use of mobile data, 54
Hardy, Seth, 216
Harris, Shane, 65
Harvey, Nick, 181
hashtag bot-flooding, 162
Hayden, Michael, 175, 189
HBGary, 65
Hengchun earthquake, 45
Hewlett-Packard
    as big-data market leader, 267n55
    as biggest PC maker, 79
    OS security vulnerabilities, 230
Hezbollah, 158
homosexuality, cyber controls over content, 39, 96, 171
Hormuud Telecom, 82, 85, 86-87
Huang, Andrew, 230
Huawei, 77, 79
humanitarian aid, enabled through big data, 54-55
Human Rights Watch, on anti-terror laws, 125-26
Hungary, human rights non-compliance, 206
Hurricane Sandy, 43
Hyppönen, Mikko, 206

IBM
    as big-data market leader, 267n55
    on data creation explosion, 52
IGF (Internet Governance Forum), 69-71, 80
images, used by ISPs, 58
India
    cyber espionage against, 23, 24, 28, 149, 150-53
    cyber espionage by, 148
    cyber filtering and surveillance, 19, 40, 93-94, 109, 122, 276n94
    cyberspace challenges and solutions, 92-94, 95
    Internet cable system, 45
    Internet disruptions, 45
    observer of Asian security alliance, 80
    "other requests" by, 114, 115

Indonesia
    cyber espionage against, 23
    cyber filtering and surveillance, 19, 39
    human rights non-compliance, 206
    requests for cyber surveillance by RIM, 122
Informatica, 66
Information Warfare Monitor Project, 262n21
In-Q-Tel, 65-66
Instagram, 269n60
Intel, 267n55
Intelligence Support Systems trade show, 208-10
International Multilateral Partnership Against Cyber Threats (IMPACT), 173-74
International Telecommunications Union (ITU), 80, 173-74
Internet
    see also cyberspace
    as distinct from cyberspace, 262n21
    as human right, 71
    censorship of, by governments, 19, 38-40, 74-75, 166, 276n94
    censorship of, not expected, 3-4
    Citizen Lab's mission to research, 5
    communications shaped by infrastructure, 5-6
    creation, 7-8, 71, 72, 232
    filters and chokepoints, 31-34, 37-38, 40-43
    in developing countries, 82, 87-90
    need for continuing absence of centralized control, 240-41
    resiliency and vulnerabilities, 43-47, 172-73
    tracking of users, 57-58
Internet Corporation for Assigned Names and Numbers (ICANN), 80, 174
Internet Engineering Task Force (IETF), 80
Internet Exchange Points (IXPs), 40-43
Internet Governance Forum (IGF), 69-71, 80
Internet Health Organization (IHO), 249
Internet service providers (ISPs)
    data collection, 41, 59
    data retention, 62-63
    filtering by, 34, 39-40, 41, 43
Iran
    attacked by Stuxnet virus, 17, 103, 176-78, 182
    cyber controls, 97
    cyber espionage against, 23
    cyber filtering and surveillance, 211, 213
    Internet cable system, 45
    observer of Asian security alliance, 80
    and Syrian Arab Spring, 155
    use of cyber crime methods, 166-67
    using cyberspace to advantage, 18

North Korea, cyber filtering, 75
Northrop Grumman, 201, 205
NSA *see* U.S. National Security Agency (NSA)
Nye, Joseph, 200

Obama, Barack
    and cyber warfare, 177, 183, 290n177
    support of FISA, 42
Oman
    cyber censorship, 19, 39-40
    cyber controls, 96
    Internet cable system, 45
Omantel, 40, 96
Ó Murchú, Liam, 178
OpenNet Initiative (ONI)
    and Belarus elections, 163-64
    book launch at IGF, 69-71
    creation of, 262n18
    and Kyrgyzstan elections, 163
    on Internet censorship, 18, 39
Operation Network Raider, 184

Pakistan
    cyber censorship, 39, 40
    cyber espionage against, 23
    emergency disabling of mobile technolo-
        gies, 94-95
    hostilities with India, 93
    Internet disruptions, 45
    observer of Asian security alliance, 80
    "other requests" by, 115
    requests for cyber surveillance by RIM, 122, 124
Pakistan Telecom, 40
Palantir, 65-66
Palfrey, John, 262n18
Parsons, Christopher, 129
Passport Canada, 113, 114, 116
passports, accessibility of, 142
Pastebin, 222
Path, 60, 269n60
PayPal, 120-21
pay-per-click schemes, 134
pay-per-install schemes, 134
PBS, 188
Personal Information Protection and
    Electronic Documents Act (PIPEDA), 118
Petraeus, David, 66
PetroChina, 78
Philippines, cyber espionage against, 23
PittPatt, 67
Poindexter, John, 64-65
Poison Ivy trojan horse, 146, 284n146
Poland, data retention, 63

policing of cyberspace
    by private sector, 129-30
    and cyber crime, 127-29
    extrajudicial, 117-20, 128-29
    government demands vs. public rights, 122,
        129, 130-31
    and national security, 124-26
    private sector requests, 121-22
    transparency reports on, 112-17
    WikiLeaks case, 120-21
Portugal, cyber espionage against, 23
Preventing Real Online Threats to Economic
    Creativity and Theft of Intellectual
    Property (Protect IP) Act (U.S.), 106,
    278n106
privacy
    and anti-terrorism laws, 125
    and biometrics, 67-68
    and cybercrime laws, 127-29
    lack of, in cloud computing, 14
    lack of, in social network services, 56-58,
        59-61
    and "other requests", 114, 117-20, 127-30
    and private-sector policing of data, 129-30
    risks to, 35, 36-37, 63-64, 130-32
Privacy International, 196, 211
private sector
    and copyright enforcement, 110
    policing of cyberspace, 129-30
    power over cyberspace, 104-10
    requests for policing, 121-22
    response to cyber crime, 140
    role in distributed stewardship of
        cyberspace, 239, 241, 242
Protecting Children from Internet Predators
    Act, 128
Psiphon (tool), 75
Public Safety Canada, 129
"push" vs. search, 36
Putin, Vladimir, 91

al-Qaeda, 4
Qatar
    cyber filtering and surveillance, 19
    Internet cable system, 45
    Internet disruptions, 45
Qin Gang, 27

Raab, Dominic, 128
Raytheon, 205
RCMP
    and Citizen Lab's Koobface findings, 138-40
    requests for user data, 129

universities, role in stewardship of cyberspace, 244

U.S. CIA, 65

U.S. Electronic Privacy Information Center (EPIC), 62

user agreements
restrictiveness of, 230-31
and use of personal data, 37, 58

U.S. FBI, 188, 218, 221

U.S. Federal Trade Commission, 61, 261n61

Ushahidi platform, 54-55

U.S. National Security Agency (NSA)
CSEC's U.S. counterpart, 1, 252n1
review of Google security, 61-62, 262n61
surveillance of citizen communications, 41-42
surveillance using satellites, 47-48
surveillance using undersea cables, 46
training in cyber intelligence, 211

U.S. Patriot Act, 14

Uzbekistan
code-of-conduct proposal, 174
cyber controls, 91-92
cyber surveillance, 213
in Asian security alliance, 80

VASTech, 212

Venezuela, cyber filtering and surveillance, 19

Verint Systems Inc., 66

Verizon
EFF lawsuit against, 42
retention of user data, 62
security breach, 188

Vietnam
human rights non-compliance, 206
Internet user growth rate, 89, 267n89
using cyberspace to advantage, 18

Viewdle, 67

Villasenor, John, 63

Villeneuve, Nart
Burmese DDOS investigation, 166
GhostNet investigation, 21, 24
Koobface investigation, 273n133
Shadow Network investigation, 149
TOM-Skype investigation, 119

VirusBlokAda, 176

viruses
as type of malware, 31
and cyber warfare, 78
Stuxnet, 17, 103

Visa, 120

Voice of America, 166-67

VUPEN Security, 201, 206

Wales, Jimmy, 127

Walton, Greg, 21-22, 25

Wang Xiaoning, 119

Websense, 19

Western Union, and WikiLeaks donation processing, 120

Wikibon, 56

WikiLeaks
as social movement, 8
information on DigiTask, 204
link with Anonymous, 222
plans to attack, 65
release of "Spy Files," 196, 200, 206
and U.S. State Department cables, 120-21

Wireshark, 34-35

worms
as type of malware, 31
Koobface, 122-41

Wuergler, Mark, 34

Wu Hongbo, 80

Wu, Tim, 271n73

Yahoo!
balance of human rights and local laws, 108
and Indian cyber surveillance, 94, 109
and requests for data or removal of content, 119
user data, 58

Yellow/Atlantic Crossing-2 cable system, 45

Yelp, 269n60

Yemen
cyber controls, 96
cyber filtering and surveillance, 211

Y.net, 96

York, Jillian, 107

youth, as early adopters of technology, 89

YouTube
Brazilian censorship of, 109
Pakistani censorship of, 39, 40
removal requests turned down, 113
Thai censorship of, 39, 97
Turkish censorship of, 95, 276n95
use by drug cartels, 98-99

ZAAD (Somalia), 82

Zafirovski, Mike, 77

zero day software
costs of, 205
defined, 142
examples of, 197, 205

Zittrain, Jonathan, 219, 229, 249, 254n18

ZTE, 77, 212